机械工程英语词汇手册

Mechanical Engineering English Vocabulary

主　审　舒启林
主　编　常　乐
参　编　李东阳　闵益杰　王福杰
　　　　汪柏良　罗荐文

北京理工大学出版社
BEIJING INSTITUTE OF TECHNOLOGY PRESS

版权专有　侵权必究

图书在版编目（CIP）数据

机械工程英语词汇手册 / 常乐主编. --北京：北京理工大学出版社，2024.5
　　ISBN 978-7-5763-3999-4

Ⅰ.①机… Ⅱ.①常… Ⅲ.①机械工程-英语-词汇-手册 Ⅳ.①TH-62

中国国家版本馆 CIP 数据核字（2024）第 098750 号

责任编辑：王梦春		**文案编辑**：芈 岚	
责任校对：刘亚男		**责任印制**：李志强	

出版发行 / 北京理工大学出版社有限责任公司
社　　址 / 北京市丰台区四合庄路 6 号
邮　　编 / 100070
电　　话 /（010）68914026（教材售后服务热线）
　　　　　（010）68944437（课件资源服务热线）
网　　址 / http://www.bitpress.com.cn

版 印 次 / 2024 年 5 月第 1 版第 1 次印刷
印　　刷 / 三河市华骏印务包装有限公司
开　　本 / 710 mm×1000 mm　1/16
印　　张 / 8.5
字　　数 / 197 千字
定　　价 / 36.00 元

图书出现印装质量问题，请拨打售后服务热线，负责调换

前　言

　　词汇是语言的构成基础。用以体现提问、描述、讨论等语篇功能的学术词汇，以及用来表达特殊专业领域的技术词汇(即专业术语)是任何一个专业学者学术能力的重要体现。近年来，伴随着机械工业在经济体系中的蓬勃发展，机械专业人才阅读专业英语文献和规范、在国际上发表最新研究或应用成果的学术需求与日俱增。然而，如果其对学术英语词汇和专业术语的了解不足，这将直接导致他们的学术交流受限。

　　近些年来，计算机检索技术正逐渐使利用语料库制定词汇表成为外语教学的主流。《机械工程英语词汇手册》(以下简称"本《手册》")正是在这样的大背景下应运而生的。经过机械工程专家推荐，在保证涵盖所有专业方向的前提下，编写组确定了 International Journal of Machine Tools and Manufacture(《机械工具和制造国际学刊》)、Journal of Mechanical Design(《机械设计学刊》)、Transactions of the ASME(《美国机械工程师学会汇刊》)等 4 本国际上权威的学术期刊，并从中选取了近 5 年由以英语为母语的研究者撰写的 120 篇学术论文，建成了 75 万余词的"机械工程学术语料库"。参照词汇的出现频率和覆盖面，编写组利用词汇检索软件 RANGE 制成了包含 37 个专业术语、361 个学术词汇的"机械工程英语词汇表"。根据词汇在语料库中出现的频率高低，学术词汇被分类为：核心学术词汇(17 个)、重要学术词汇(42 个)，以及一般学术词汇(302 个)。

　　本《手册》中的所有词汇均按照"词汇、音标、词性、词义、例句及汉语译文"的顺序呈现。为了体现词汇在机械工程英语中最地道、最常见的使用状态，本《手册》对核心和重要学术词汇从语料库中选取 3 个例句，对一般学术词汇和专业术语分别给出 2 个和 1 个例句，并将这些词及其最常见的搭配以异色的形式着重标记出来，以便引起使用者的注意。

基于"机械工程英语词汇表"和"机械工程学术语料库"编制的本《手册》着眼于机械工程专业人士的学术读写与交流需求。据编写组统计，"机械工程英语词汇表"中的398个单词在"机械工程学术语料库"中的覆盖率达到11.01%，加上占比76.1%的英语通用词汇（即2 000个基础核心词汇）以及5%～10%的专有名词、缩略词等，基本上能覆盖机械学术语篇中95%的内容。换言之，在熟悉英语中常见的2 000个通用词的基础之上，掌握本《手册》中的398个词汇可以有效帮助机械专业学习者和研究人员实现无障碍阅读本专业英语学术文本。因此，本《手册》能满足机械工程专业研究生和高年级本科生的学术英语学习需求，并能助力中国机械学者和技术人员使用英语与国际机械学术界接轨。

本《手册》由沈阳理工大学外国语学院"理工科学术英语教学团队"负责人常乐教授和英语笔译专业研究生李东阳、闵益杰、王福杰、罗荞文，以及机械工程专业研究生汪柏良共同编写，并由机械工程专家舒启林教授全程进行专业指导和审阅。本《手册》作为2019年度中国外语教材研究专项课题的前期成果，获得北京外国语大学中国外语教育研究中心许家金教授的悉心指导，同时还获得教育部产学合作协同育人项目"基于语料库的英汉对照机械工程学术词汇与专业术语表创建研究"和2020年辽宁省教育厅科研经费项目"基于语料库的机械英语词汇表创建"的资助；前期的语料库和词汇表制作过程获得沈阳建筑大学外国语学院吴明海、陈颖两位副教授的技术支持，编者在此一并致以衷心的谢意！

鉴于编者水平有限，书中难免有不足与疏漏之处，敬请广大使用者批评与指正。

本《手册》编写组于沈阳理工大学

目 录

第一部分　专业术语 ·· (1)
　　词汇表 ··· (1)
　　词汇详解 ··· (3)

第二部分　核心学术词汇 ··· (9)
　　词汇表 ··· (9)
　　词汇详解 ·· (10)

第三部分　重要学术词汇 ·· (16)
　　词汇表 ·· (16)
　　词汇详解 ·· (18)

第四部分　一般学术词汇 ·· (33)
　　词汇表 ·· (33)
　　词汇详解 ·· (46)

References（参考文献） ·· (126)

第一部分 专业术语

词汇表

单词	音标	词性	词义
aluminium	[ˌæləˈmɪnɪəm]	n.	铝
amplitude	[ˈæmplɪtjuːd]	n.	振幅
axial	[ˈæksɪəl]	adj.	轴的；轴向的
axis	[ˈæksɪs]	n.	轴
chip	[tʃɪp]	n.	切屑
coefficient	[ˌkəʊɪˈfɪʃnt]	n.	系数
coupling	[ˈkʌplɪŋ]	n.	耦合
damping	[ˈdæmpɪŋ]	n.	阻尼
diagram	[ˈdaɪəgræm]	n.	图表
dynamics	[daɪˈnæmɪks]	n.	动力学
electrode	[ɪˈlektrəʊd]	n.	电极
feed	[fiːd]	n.	进给
flexible	[ˈfleksəb(ə)l]	adj.	柔性的
fluid	[ˈfluːɪd]	n.	流体，液体
friction	[ˈfrɪkʃn]	n.	摩擦
hydraulic	[haɪˈdrɒlɪk]	adj.	液压的
interface	[ˈɪntəfeɪs]	n.	界面

续表

单词	音标	词性	词义
jet	[dʒet]	n.	射流
laser	[ˈleɪzə(r)]	n.	激光
lens	[lenz]	n.	透镜
loop	[luːp]	n.	环，回路
mathematical	[ˌmæθəˈmætɪkəl]	adj.	数学的
microscope	[ˈmaɪkrəskəʊp]	n.	显微镜
mode	[məʊd]	n.	模式
modulus	[ˈmɒdjʊləs]	n.	模量
nozzle	[ˈnɒzəl]	n.	喷嘴
output	[ˈaʊtpʊt]	n.	输出
phase	[feɪz]	n.	相位
pitch	[pɪtʃ]	n.	螺距
radius	[ˈreɪdiəs]	n.	半径
segment	[ˈseɡmənt]	n.	段
sensor	[ˈsensə(r)]	n.	传感器
shear	[ʃɪə(r)]	n.	剪切
trajectory	[trəˈdʒektəri]	n.	轨迹
vector	[ˈvektə(r)]	n.	向量，矢量
velocity	[vəˈlɒsəti]	n.	速度
voltage	[ˈvəʊltɪdʒ]	n.	电压

词汇详解

1. aluminium ［ˌæləˈmɪniəm］ n. 铝

例句 Special procedures were proposed in the past to maximise edge retention of the specimens using nickel and **aluminium** oxide filler in a vacuum cast epoxy mount.

（人们）在过去曾提出过特殊的程序，通过在真空铸造环氧贴片中使用镍和氧化铝填充物来最大限度地保持试样的边缘。

2. amplitude ［ˈæmplɪtjuːd］ n. 振幅

例句 The **amplitude** at the chatter frequency is highest.
振幅在颤振频率处达到最大。

3. axial ［ˈæksiəl］ adj. 轴的；轴向的

例句 Resampling of the tool geometry is done by interpolating at each **axial** segment.

刀具几何参数的重新采样是通过在每个轴段内插值完成的。

4. axis ［ˈæksɪs］ n. 轴

例句 Since the magnitude response of the **axis** controllers drops off at those frequencies, a simple servo system creates actual circles that are smaller than the reference path.

由于轴控制器的幅值响应在这些频率处下降，因此一个简单的伺服系统会生成的实际圆要小于参考路径。

5. chip ［tʃɪp］ n. 切屑

例句 The overall uncut **chip** thickness is generated by the axial rigid body motion of the drill (feed motion) and by vibrations of the drill in lateral, axial and torsional directions.

整体毛边厚度由钻头的轴向刚体运动（进给运动）以及钻头在横向、轴向和扭转方向上的振动产生。

6. coefficient ［ˌkəʊəˈfɪʃənt］ n. 系数

例句 The friction **coefficient** of 0.15 was selected for the Coulomb friction model for all the contact surfaces.

所有接触面的库仑摩擦模型选择的摩擦系数都是 0.15。

7. coupling [ˈkʌplɪŋ] n. 耦合

例句 Much research has been directed towards the calculation of path error in efficient and realistic ways, and to determining the optimal cross **coupling** control action.

许多研究都是针对路径误差的有效和现实方式的计算，并确定最优交叉耦合控制行为。

8. damping [ˈdæmpɪŋ] n. 阻尼

例句 Anchor **damping** can also arise from imperfections in the fabrication of the device, which could cause slight imbalances of the vibrational mode.

锚杆阻尼也可能来自设备制造中的缺陷，它可能会导致振动模式的轻微失衡。

9. diagram [ˈdaɪəɡræm] n. 图表

例句 The block **diagram** of this control system is shown in Fig. 1.

此控制系统的框图如图 1 所示。

10. dynamics [daɪˈnæmɪks] n. 动力学

例句 The beam theory was used to model the **dynamics** of micro-drills.

采用梁理论对微钻头进行动力学建模。

11. electrode [ɪˈlektrəʊd] n. 电极

例句 This is because small bubbles will attach to the **electrode** until departure size has been reached.

这是因为小气泡会附着在电极上，直至达到脱落的尺寸。

12. feed [fiːd] n. 进给

例句 In this case, velocity saturation can only be avoided if the **feed rate** is actively reduced during the phases of high velocity.

在这种情况下，只有在高速阶段主动降低进给速率才能避免速度饱和。

13. flexible [ˈfleksəb(ə)l] adj. 柔性的

例句 Stability validation has been carried by a series of orbital drilling tests with a

flexible tool and rigid block of work material.

采用柔性工具和刚性工作块材料开展了一系列螺旋钻孔试验来验证稳定性。

14. fluid ['fluːɪd] n. 流体，液体

例句 These materials are composed of a functional **fluid** that contains a suspension of metallic particles.

这些材料由包含金属颗粒悬浮液的功能性流体组成。

15. friction ['frɪkʃn] n. 摩擦

例句 The drive inertia and viscous **friction** parameters are identified using linear least square technique.

采用线性最小二乘法对驱动惯量和黏性摩擦参数进行辨识。

16. hydraulic [haɪ'drɒlɪk] adj. 液压的

例句 Electrostatic **micro-hydraulic** structures can be utilized for sensing applications as well.

静电微型液压结构也可用于传感应用。

17. interface ['ɪntəfeɪs] n. 界面

例句 On the contrary, if the acceleration is in reverse direction, the **interface** will become unstable and interface will no longer be a plane.

反之，如果加速度方向相反，则界面会变得不稳定且不再是平面。

18. jet [dʒet] n. 射流

例句 Due to the rising centre feature, the stagnation point at the centre of the **jet** is now exacerbated.

中心上升的特点加剧了射流中心的驻点。

19. laser ['leɪzə(r)] n. 激光

例句 Melt pool shape tends to be more asymmetric with increasing laser **power** and increasing scan velocity, especially for Type Ⅱ melt pools.

随着激光功率和扫描速度的增大，熔池形状变得更加不对称，尤其是Ⅱ型熔池。

20. lens [lenz] n. 透镜

例句 One of the biggest problems is the existence of the **lens** glare around the melt pool region.

一个最大的问题是在熔体池区域周围存在透镜眩光。

21. loop [luːp] n. 环

例句 Open-**loop** frequency responses are then algebraically computed from the **closed-loop** frequency responses.

然后通过闭环频率响应代数来计算开环频率响应。

22. mathematical [ˌmæθəˈmætɪkəl] adj. 数学的

例句 In order to obtain the position errors, a **mathematical** model is proposed.

为了得到位置误差，提出了一个数学模型。

23. microscope [ˈmaɪkrəskəʊp] n. 显微镜

例句 In all three cases, feedback control was performed using an **optical microscope**, which limits positioning accuracy and prevents the use of transparent controllers.

这三种情况下，都使用光学显微镜进行反馈控制会限制定位精度并阻止使用透明控制器。

24. mode [məʊd] n. 模式

例句 This wear **mode** was initiated upon adhesion of work-piece material onto the tool surface near the cutting edges.

这种磨损模式是在工件材料黏附在靠近切削刃的刀具表面时开始的。

25. modulus [ˈmɒdjʊləs] n. 模量

例句 Its large Young's **modulus** and low friction coefficients make SiC an ideal material for smart implants and in vivo biosensors.

其较大的杨氏模量和较低的摩擦系数使碳化硅成为智能植入和体内生物传感器的理想材料。

26. nozzle [ˈnɒzəl] n. 喷嘴

例句 Instead, a new parameter is required which defines the spacing between **nozzle** rim contour and the work.

相反，需要一个新的参数来定义喷嘴边缘轮廓和工件之间的间距。

27. output ['aʊtpʊt] n. 输出

例句 The **output** from the tool-foil thermocouple method was a voltage measurement of the tool-foil circuit.

工具箔热电偶法中的输出是工具箔电路的电压测量值。

28. phase [feɪz] n. 相位

例句 These systems generally offer less stiffness than authentic **phase-change** systems, but they respond to the phase-change energy much more quickly.

这些系统通常比真正的相变系统提供更小的刚度，但它们对相变能量的响应要快得多。

29. pitch [pɪtʃ] n. 螺距

例句 The actual **pitch** angle additionally depends on the axial run-out and feedrate.

实际螺距角还取决于轴向跳动和进给速度。

30. radius ['reɪdɪəs] n. 半径

例句 The average edge **radius of** the tool was calculated to be 3mm from SEM images (Fig. 1).

根据 SEM 图像（见图 1），计算出刀具的刀尖圆角半径为 3 mm。

31. segment ['seɡmənt] n. 段

例句 For this, the length of modified linear **segment** and inserted spline **segment** is first calculated.

为此，首先计算修正后的直线段和插入的样条线段的长度。

32. sensor ['sensə(r)] n. 传感器

例句 In the future, this constraint can be loosened by limiting the noise of both the **sensor** and interface circuit, while improving the quality factor of the resonator.

未来，可以通过限制传感器和接口电路的噪声以及提高谐振器的质量因数，来减少这种约束。

33. shear ［ʃɪə(r)］ n. 剪切

例句　In the cutting process, parameters like **shear angle**, average friction angle and shear yield stress are difficult or impossible to measure.

在切削过程中很难或者不可能测量到剪切角、平均摩擦角、剪切屈服应力等参数。

34. trajectory ［trəˈdʒektəri］ n. 轨迹

例句　To solve this problem, Weck suggested filtering out the high frequency components of the reference **trajectory**.

为了解决这个问题，Weck 建议滤掉参考轨迹的高频分量。

35. vector ［ˈvektə(r)］ n. 向量，矢量

例句　The non-periodic knot **vector** is constructed as to ensure symmetry across the angular bisector of the corner angle.

构造非周期节点向量是为了确保在角平分线上的对称性。

36. velocity ［vəˈlɒsəti］ n. 速度

例句　A high frequency switching controller is then used to adjust the overall **path velocity** to accommodate the needs of all axes.

然后，使用高频开关控制器来调整总体路径速度，以适应所有轴的需求。

37. voltage ［ˈvəʊltɪdʒ］ n. 电压

例句　As the **voltage** increases, the membrane moves exert a downward force on the rubber tip, which is measured by the scale.

随着电压的增加，薄膜的移动对橡胶尖端施加一个向下的力，这个力可由量表测量出来。

第二部分 核心学术词汇

词汇表

单词	音标	词性	词义
analysis	[əˈnælɪsɪs]	n.	分析
approach	[əˈprəʊtʃ]	n.	方法
area	[ˈeəriə]	n.	领域，面积，区域
contact	[ˈkɒntækt]	n.	接触
data	[ˈdeɪtə]	n.	数据
design	[dɪˈzaɪn]	n.	设计
error	[ˈerə]	n.	误差
function	[fʌŋkʃn]	n.	函数
maximum	[ˈmæksɪməm]	adj.	最大的
method	[ˈmeθəd]	n.	方法
parameters	[pəˈræmɪtəz]	n.	参数（复数）
process	[ˈprəʊses]	n.	流程，工艺流程
required	[rɪˈkwaɪəd]	v.	要求，需要（过去分词）
section	[ˈsekʃn]	n.	章节；部分；区域
significant	[sɪgˈnɪfɪkənt]	adj.	显著的，大量的
similar	[ˈsɪmələ(r)]	adj.	类似的，相似的
structure	[ˈstrʌktʃə(r)]	n.	结构

词汇详解

1. analysis [əˈnælɪsɪs] n. 分析

例句 1 Non-integer values are relevant to the **analysis** of dynamic instability.
非整数值与动态不稳定性的分析相关。

例句 2 A mathematical framework of **reliability analysis** models these uncertainties as random variables.
可靠性分析的数学框架将这些不确定性建模为随机变量。

例句 3 This product had 576 reviews on Amazon.com and, of those, 50 reviews were chosen for the **manual analysis**.
这个产品在亚马逊网站上有576条评论，选其中的50条进行人工分析。

2. approach [əˈprəʊtʃ] n. 方法

例句 1 Our **approach to** quantify the shape of the Pareto front is discussed next.
接下来讨论我们用来量化帕累托解集合形状的方法。

例句 2 Finally, experimental tests to validate this **approach to** the inverse problem have been performed.
最后，通过实验测试验证了该逆向问题方法的有效性。

例句 3 A common **approach for** determining the counterfactual is taking measurements from a control group concurrently with the measurements of the impacted group.
确定反事实的一种常见方法是对受影响组和对照组进行同时测量。

3. area [ˈeəriə] n. 领域，面积，区域

例句 1 The volume to **surface area** ratio is used as a measure of geometric complexity.
体积与表面积的比率是用来衡量几何体复杂性的指标。

例句 2 For ductile materials, it is also important to reduce the **contact area** between the grit and workpiece.
对于韧性材料，减小磨粒与工件之间的接触面积也很重要。

例句 3 Regarding the analysis of eye-tracking data, we manually created **area of interest**（AOI）for each sustainable attribute and the overall text area of stimulus in task

4, as shown in Fig. 8.

关于眼球追踪数据的分析,如图8所示,我们为任务4中的每个可持续属性和刺激物的整体文本区域手动创建了兴趣区(AOI)。

4. contact [ˈkɒntækt] n. 接触

例句1 The tool tip came **in contact with** the copper foil intermittently and generated voltage pulses in the tool-foil circuit [Fig. 2(b)].

刀尖与铜箔间歇性地接触,在刀箔电路中产生电压脉冲[见图2(b)]。

例句2 Indeed, it has opened the door for many applications of wetting phenomena, capillary forces, **contact angle** formation, etc.

事实上,它使润湿现象、毛细力、接触角形成等方面的许多应用成为可能。

例句3 On the other hand, the **contact forces** generated by the bottom of the end mill are neglected in the proposed model.

另一方面,该模型忽略了立铣刀底部所产生的接触力。

5. data [ˈdeɪtə] n. 数据

例句1 This yields 66 **data points** per tool revolution for the fastest spindle speed (4500 r/min).

最快主轴转速(4500 r/min)刀具每旋转一周会产生66个数据点。

例句2 The user wants to determine the calibration parameters such that the model prediction is consistent with **experimental data**.

用户需要确定标定参数,以使模型的预测与实验数据一致。

例句3 The analysis of the **measured data** allowed a clear understanding of the machine's thermal behavior, by relating the volumetric error variation to the machine's activity.

通过对测量数据的分析,将体积误差变化与机器的活动联系起来,可以清楚地了解机器的热行为。

6. design [dɪˈzaɪn] n. 设计

例句1 This can speed up time in the **design process** and move the product toward completion.

这可以加速设计过程,推动产品完成。

例句2 One way to shorten the **design process** can be through the use of additive manufacturing.

缩短设计过程的一种方法是使用增材制造。

例句3 If the bounds on the **design variables** are too small, we might miss feasible designs.

如果设计变量的界限太小,我们可能会错过可行的设计。

7. error [ˈɛrə] n. 误差

例句1 The **tracking error** between the desired trajectory and the real one is reduced and the etched profile is improved.

这样能减小预定轨迹与真实轨迹之间的跟踪误差,改善蚀刻轮廓。

例句2 However it is possible to bound the **error of** numerical integration between two given limits, s_{min} and s_{max}.

然而,有可能将数值积分的误差限定在两个给定的极限 s_{min} 和 s_{max} 之间。

例句3 The value of length l is evaluated by imposing a user-defined **position error** tolerance limit constraint.

长度 l 的值是通过施加一个用户定义的位置误差容限约束来评估的。

8. function [fʌŋkʃn] n. 函数

例句1 Figs. 4-8 show the force model coefficients as a **function of** cutting distance.

图4至图8显示了力模型系数与切削距离之间的函数关系。

例句2 Local optimization is performed in each subregion by using the quadratic model as the **objective function**.

通过使用二次模型作为目标函数,可在每个子区域进行局部优化。

例句3 The addition of these multivariate samples produces a refined surrogate that is capable of capturing the variate interactions in the **performance function**.

加入这些多变量样本后,会产生一个更优替代品,它能够捕捉到性能函数中的变量交互作用。

9. maximum [ˈmæksɪməm] adj. 最大的

例句1 To address this, the second term in Eq. (7) is set to zero when the value of q reaches the **maximum value**.

为了解决此问题,当 q 值达到最大时,公式(7)中的第二项被设为0。

例句 2　Moreover, **maximum temperature**, melt pool length and width decrease with increasing speed function.

此外，最高温度、熔池长度和宽度随着速度函数的增加而减少。

例句 3　The **maximum speed** of each axis drive is 200 mm/s and the maximum acceleration is approximately 10 m/s^2.

每个轴驱动器的最大速度为 200 mm/s，最大加速度约为 10 m/s^2。

10. method　['mɛθəd]　n. 方法

例句 1　Another advantage of the **proposed method** lies in its simplicity in fixtures.

所提方法的另一优点在于它的夹具简单。

例句 2　A **method to** achieve this goal by modulating axis speed in concert with error has been presented by Seethaler and Yellowley.

Seethaler 和 Yellowley 提出了一种通过误差调节轴速来实现这一目标的方法。

例句 3　The paper presents a **method for** selecting grinding conditions and assists researchers to understand the complex dynamics of centreless grinding.

本文提出了一种选择磨削条件的方法，并协助研究人员了解无心磨削的复杂动力学。

11. parameters　[pəˈræmɪtəz]　n. 参数（复数）

例句 1　Equation (11) may be written in terms of geometry **parameters of** the gear.

可以用齿轮的几何参数表示公式(11)。

例句 2　The resulting schematics can be used to inform selection of **parameters for** given applications.

由此产生的示意图可用于为特定应用的参数选择提供信息。

例句 3　The **parameters in** the equations are given in Table A.2 and are used in the analysis of the models in this section.

表 A.2 给出了方程中的参数，并用于本节的模型分析。

12. process　['prəʊses]　n. 流程，工艺流程

例句 1　The **process of** attaining the vacuum environment in the furnace requires multiple steps.

炉内达到真空环境的过程需要多个步骤。

例句 2　It should be noted that this approach has been extended to include **process**

parameters such as force, chip thickness, and so on.

需要注意的是，这个方法被扩展到包括力和切屑厚度等在内的工艺参数。

例句3　Therefore, proper selection of **process parameters** will help improve the mechanical property along with decreasing the surface roughness.

因此，合理选择工艺参数有助于提高机械性能，同时降低表面粗糙度。

13. required ［rɪˈkwaɪəd］ v. 要求，需要(过去分词)

例句1　The students were **required to** use bio-inspired design to address the design problem and had some exposure to the field.

要求学生们使用仿生设计来解决设计问题，并对该领域有了一定的了解。

例句2　Hence, accurate object positioning would be difficult to achieve, which is **required in** applications such as assembly tasks.

因此，在诸如装配作业之类的应用中所必需的精确目标定位难以实现。

例句3　Since no manual analysis is **required for** its calculation, the WPR(weighted phrase rating) can be calculated from all available reviews very quickly.

由于该计算无需人工分析，因此可从所有可获取的评论中非常迅速地计算出加权短语评级(WPR)。

14. section ［ˈsekʃn］ n. 章节；部分；区域

例句1　As discussed in the **previous section**, the print quality is best when the feed direction aligns with the scan velocity.

正如上一节所讨论的，当进纸方向与扫描速度一致时，打印质量最好。

例句2　The highlighted region in Fig. 3 (a) shows the **cross section** of the overall extruded material.

图3(a)中的高亮区域显示了整个挤压材料的横截面。

例句3　In **this section**, we introduce the force model coefficients, describe the methods of estimating them in real-time, and then show how they are insensitive to the cutting conditions.

在本节中，我们将介绍力模型系数，描述实时估算它们的方法，然后展示它们如何对切削条件不敏感。

15. significant ［sɪgˈnɪfɪkənt］ adj. 显著的，大量的

例句1　This method provided **significant increase** of 200% in fabrication yield for our devices.

这种方法使我们设备的制造产量大幅提高了200％。

例句2　However, it still requires **a significant amount of** manual reading which is very time-consuming.

然而，它仍然需要大量的人工阅读，非常耗费时间。

例句3　The ratio between the tool length and diameter has been shown to **have a significant impact on** surface roughness of a part.

刀具长度和直径之间的比率对零件的表面粗糙度有很大影响。

16. similar　　［ˈsɪmələ(r)］　　adj. 类似的，相似的

例句1　It is expected that the data mining analysis will produce **similar results**.

预计数据挖掘分析将产生相似的结果。

例句2　**Similar to** cutting forces, torque and power requirement by tool and workpiece spindles must be calculated.

与切削力类似，必须要计算出刀具及工件主轴的扭矩和功率要求。

例句3　It is observed that the profile along the transverse direction shows **similar trend** in all the cases.

可以看出，在所有情况下，沿横向的剖面显示出类似的趋势。

17. structure　　［ˈstrʌktʃə(r)］　　n. 结构

例句1　The **overall structure** of the position weighted transition gain is shown in Fig. 5.

位置加权过渡增益的整体结构如图5所示。

例句2　Section 2 introduces the **structure of** the combined driver model and desired path generation.

第2节介绍了组合驱动模型的结构和期望路径的生成。

例句3　Due to its L-shape, **test structure** 2A exhibits mass asymmetries in both lateral axis x and y.

由于测试结构2A呈L形，该结构在x轴和y轴上均表现出质量不对称。

第三部分　重要学术词汇

词汇表

单词	音标	词性	词义
accuracy	[ˈækjuərəsi]	n.	准确性
achieve	[əˈtʃiːv]	v.	实现
algorithm	[ˈælɡərɪðəm]	n.	算法
assembly	[əˈsembli]	n.	装配，组装
components	[kəmˈpəʊnənts]	n.	分量；部件(复数)
constant	[ˈkɒnstənt]	adj.	常存在的，恒定的，不变的
conventional	[kənˈvenʃənl]	adj.	传统的
corresponding	[ˌkɒrəˈspɒndɪŋ]	adj.	相应的，对应的
defined	[dɪˈfaɪnd]	v.	定义，明确(过去分词)
diameter	[daɪˈæmɪtə(r)]	n.	直径
elements	[ˈelɪmənts]	n.	要素(复数)
energy	[ˈenədʒi]	n.	能量
equation	[ɪˈkweɪʒn]	n.	公式
equations	[ɪˈkweɪʒnz]	n.	方程(复数)
errors	[ˈerərz]	n.	误差(复数)
factor	[ˈfæktə(r)]	n.	系数，因素
final	[ˈfaɪnl]	adj.	最终的
generated	[ˈdʒenəˌreɪtɪd]	v.	产生(过去分词)

续表

单词	音标	词性	词义
individual	[ˌɪndɪˈvɪdʒuəl]	adj.	单独的，个人的
initial	[ɪˈnɪʃl]	adj.	初始的
input	[ˈɪnpʊt]	n.	输入
linear	[ˈlɪniə(r)]	adj.	线性的
matrix	[ˈmeɪtrɪks]	n.	矩阵
mechanism	[ˈmekənɪzəm]	n.	机制，机构
mechanisms	[ˈmekənɪzəmz]	n.	机制(复数)
methods	[ˈmeθədz]	n.	方法(复数)
minimum	[ˈmɪnɪməm]	adj.	最小的，最低限度的
negative	[ˈneɡətɪv]	adj.	负的，负面的
obtained	[əbˈteɪnd]	v.	获得，得到(过去分词)
parameter	[pəˈræmɪtə(r)]	n.	参数
previous	[ˈpriːviəs]	adj.	以前的，先前的
range	[reɪndʒ]	n./v.	范围/从……到……
ratio	[ˈreɪʃiəʊ]	n.	比率
reference	[ˈrefrəns]	n.	参考
research	[rɪˈsɜːtʃ]	n.	研究
significantly	[sɪɡˈnɪfɪkəntli]	adv.	明显地
simulation	[ˌsɪmjuˈleɪʃn]	n.	仿真，模拟(实验)
source	[sɔːs]	n.	来源
technique	[tekˈniːk]	n.	技术
techniques	[tekˈniːks]	n.	技术(复数)
torque	[tɔːk]	n.	扭矩
volume	[ˈvɒljuːm]	n.	体积，容量

词汇详解

1. accuracy [ˈækjuərəsi] n. 准确性

例句1 The **accuracy of** the gaps can be evaluated by measuring the gaps with calipers and comparing to the intended gap width.

间隙的精度可以通过用卡尺测量并与预期间隙宽度比较来评估。

例句2 Hence the **accuracy in** strain measurement depends on the actual deformation behavior of the specimen system.

因此，应变测量的准确性取决于试样系统的实际变形行为。

例句3 A single decision tree achieved an **accuracy of** 68%, but this number varied significantly depending on the training and test data split.

单个决策树达到了68%的准确率，但这个数字因训练和测试数据的划分而有很大的差异。

2. achieve [əˈtʃiːv] v. 实现

例句1 Finally, objectives must be optimized to **achieve** the desired behavior.

最后，必须对目标进行优化，以实现预期行为。

例句2 A series of grinding tests are conducted to **achieve** the best cutting edge quality and grinding performance.

为了达到最佳的切削刃质量和磨削性能，进行了一系列的磨削试验。

例句3 In general terms, the MR fluid brake must be twice the size of the frictional disk brake in order to **achieve** the same braking torque.

一般来说，磁流变液流体制动器的大小必须是摩擦盘式制动器的两倍，才能达到相同的制动扭矩。

3. algorithm [ˈælɡərɪðəm] n. 算法

例句1 In Ref. [25] they explain their **algorithm for** a 3-axis machine.

在参考文献[25]中，他们解释了对三轴机床的算法。

例句2 We develop an **algorithm in** Python environment to execute this optimization.

我们在Python环境中开发了一种算法来执行这种优化。

例句 3　A new **algorithm** to extract the system matrix of the electrical-mechanical coupled array is introduced.

介绍了一种提取机电耦合阵列系统矩阵的新算法。

4. assembly　[əˈsembli]　n. 装配，组装

例句 1　Various stages of the **assembly process** are shown in Fig. 6.

装配过程的各个阶段如图 6 所示。

例句 2　The **assembly process** is carried out by sequential transformation of differential elements as follows.

如下文所示，装配过程是通过不同组件的顺序转换进行的。

例句 3　The actuator is an **assembly of** two dies with a plastic spacer sandwiched in between.

执行机构是由两个模具组成的组件，中间夹着一个塑料垫片。

5. components　[kəmˈpəʊnənts]　n. 分量；部件（复数）

例句 1　Fig. 24 shows normal **components of** acceleration vectors, resulting from each tool path motion.

图 24 显示了每个刀具路径运动产生的加速度矢量的法向分量。

例句 2　Fig. 11 shows the x, y and z **components of** the error at these six different poses of the machine.

图 11 展示了在这 6 个不同机器位置上误差的 x、y 和 z 分量。

例句 3　This ability enables the processing of **components with** a broad range of mechanical properties in the as-built condition.

这种能力能够在完工条件下处理具有广泛机械性能的组件。

6. constant　[ˈkɒnstənt]　adj. 常存在的，恒定的，不变的

例句 1　The following **constant parameters** are considered：$P=138$ MPa, $ma=0.03$ kg/min, SOD (stand-off distance) $= 3$ mm.

考虑以下恒定参数：$P=138$ MPa, $ma=0.03$ kg/min, SOD（相隔距离）$= 3$ mm。

例句 2　This method requires at least two data points and assumes that the **constant value** is utilized to consider the discrepancy.

这种方法需要至少两个数据点，并假设利用常数值来考虑差异。

例句 3　When the axis controllers are mismatched, the simple servo systems and

FMI(frequency modulated interpolation) result in large steady state errors along the **constant velocity** segments.

当轴控制器不匹配时，简单的伺服系统和调频插值（FMI）会导致沿恒速段出现较大的稳态误差。

7. conventional ［kənˈvenʃənl］ adj. 传统的

例句 1 In contrast, additive manufacturing (AM) techniques do not have the same extent of design constraints that limit **conventional processes**.

相比之下，增材制造（AM）技术没有限制传统工艺的相同程度的设计约束。

例句 2 However, simple-geometry, highly standardized, or massively produced parts are still cost-competitive by using **conventional machining**.

然而，简单几何形状、高度标准化或大批量生产的零件通过使用传统加工仍然具有成本竞争力。

例句 3 More, future work will focus on developing methods and tools for enabling bidirectional migration of designs between additive manufacturing and **conventional manufacturing**.

而且，未来的工作将集中于开发方法和工具，以实现增材制造和传统制造之间的设计双向迁移。

8. corresponding ［ˌkɒrəˈspɒndɪŋ］ adj. 相应的，对应的

例句 1 **Corresponding** values are listed in Table 1.

相应的数值列于表 1 中。

例句 2 The tool path **corresponding to** this part program is plotted in Fig. 1.

与此零件程序相对应的刀具路径如图 1 所示。

例句 3 Each dip is 10 min long, **corresponding to** a total development time of 20 min.

每次浸出时间为 10 min，对应 20 min 的总显影时间。

9. defined ［dɪˈfaɪnd］ v. 定义，明确（过去分词）

例句 1 The feasible design space is often **defined by** constraints.

可行的设计空间通常由约束条件来决定。

例句 2 Errors in the A-axis, shown in Fig. 3, are **defined in** a similar way.

如图 3 所示，A 轴的误差也是以类似的方式定义。

例句 3　Performance here can be **defined as** a function of weight, quality, level of product innovation, and level of customization.

这里的性能可以被定义为重量、质量、产品创新水平和定制水平的函数。

10. diameter　[daɪˈæmɪtə(r)]　n. 直径

例句 1　The **diameter of** the water orifice varies from 0.08 to 0.24 mm.

水孔的直径范围从 0.08 mm 到 0.24 mm 不等。

例句 2　Therefore, a tool length to **diameter ratio** is calculated for each facet by dividing the reachability depth by the tool diameter.

因此，可以通过刀具的可达深度除以直径来计算每个小面的刀具长径比。

例句 3　The overall **diameter of** the electrode and the gas film increased from 500 mm to 548 mm, indicating the thickness of the gas film is 24 mm.

电极和气膜的整体直径从 500 mm 增加到 548 mm，表明气膜的厚度为 24 mm。

11. elements　[ˈelɪmənts]　n. 要素(复数)

例句 1　The cutting lips are divided into finite number of cutting **elements** with oblique cutting geometry.

切削刃被划分为有限数量的具有倾斜切割几何形状的切割元素。

例句 2　In this example, the first full-system prototype was used to check the technical **elements of** the design.

在这个例子中，第一个完整的系统原型被用以检查设计的技术要素。

例句 3　These definitions emphasize prototypes' purpose as aiding in technical **elements of** the design.

这些定义强调了原型的目的在于为设计的技术要素提供帮助。

12. energy　[ˈenədʒi]　n. 能量

例句 1　Furthermore, there is no appreciable difference in behavior between Type Ⅰ and Type Ⅱ melt pool size when considering **energy density** as an all-inclusive factor.

此外，当能量密度被看作一个全面因素时，Ⅰ型和Ⅱ型熔池尺寸之间的行为没有明显的差异。

例句 2　In recent years, advanced modelling techniques for some **energy loss** mechanisms have been introduced to predict resonator performance based on fundamental physics.

近年来，引入了一些能量损失机制的先进建模技术，用以预测基于基本物理学的谐振器性能。

例句3　The SLM(selected laser melting) processing conditions which were studied are summarized in Table 2 with linear **energy densities** of 0.2-0.8 kJ/m investigated.

表2总结了所研究的选择性激光熔化(SLM)加工条件，所研究的线性能量密度为0.2~0.8kJ/m。

13. equation　[ɪˈkweɪʒn]　n. 公式

例句1　The velocity field is related to the boundary of bubble through the **equation of** continuity.

通过连续性方程将速度场与气泡边界联系起来。

例句2　The summation of different sinusoidal functions has been found as the **equation with** the best fitting.

通过对不同正弦函数的求和，得到了最佳拟合的方程。

例句3　This is then utilised to understand the effects of nozzle geometry against response profile and to demonstrate the validity of the **equation**.

然后利用这一点来了解喷嘴的几何形状对响应曲线的影响，并证明该方程的有效性。

14. equations　[ɪˈkweɪʒnz]　n. 方程(复数)

例句1　The six steps to apply Lotka-Volterra **equations in** technology evolution prediction are summarized.

总结了将Lotka-Volterra方程应用于技术演化预测的6个步骤。

例句2　The vehicle model used and all of the vehicle **equations of** motion are based on the model in [64].

使用的车辆模型和所有的车辆运动方程都是基于[64]中的模型。

例句3　Another contribution of the paper is the rigorous stability analysis of the nonlinear time-periodic **equations of** motion.

本文的另一贡献是对非线性时间周期性运动方程做出严格的稳定性分析。

15. errors　[ˈerərz]　n. 误差(复数)

例句1　**Errors in** the A-axis, shown in Fig. 3, are defined in a similar way.

图3所示的A轴误差也是以类似的方式定义的。

例句 2　New methods for measuring the thermal **errors of** rotary axes were recently introduced.

最近引进了测量旋转轴热误差的新方法。

例句 3　The simulation and experimental contour **errors for** the cases with and without constraints are summarized in Table 1.5.

表 1.5 中总结了有约束和无约束情况下的模拟及实验轮廓误差。

16. factor　[ˈfæktə(r)]　n. 系数，因素

例句 1　The **factor of** safety around 20 was good for these first articles.

安全系数在 20 左右对于第一批产品很合适。

例句 2　Experiments in section Ⅲ were undertaken in low pressure, which enabled low damping, or high **quality factor**.

第三节中的实验是在低压下进行的，可以实现低阻尼或高质量系数。

例句 3　The **quality factor** is one of the most important characteristics for a resonant system, and for most applications, a higher quality factor is desired.

品质因数是谐振系统最重要的特性之一。对于大多数应用来说，需要更高的品质因数。

17. final　[ˈfaɪnl]　adj. 最终的

例句 1　The **final design** was expected to be detailed enough that the mechanisms and biological principles are evident.

预计最终的设计将足够详细，其机制和生物原理是显而易见的。

例句 2　The data is then used to compensate the process tool-path; reducing variance in the geometry of the **final part**.

这些数据随后被用来补偿加工刀具路径；减少最终零件的几何形状差异。

例句 3　It should be advised that the **final solution** with the minimum part count may not be the optimal solution if other constraints (e.g., overhanging problem) are applied.

如果应用了其他约束条件(例如悬伸问题)，零件数最少的最终方案可能不是最优化的解决方案。

18. generated　[ˈdʒenəˌreɪtɪd]　v. 产生(过去分词)

例句 1　That means the test data are **generated by** using the true Young's modulus from the Timoshenko beam theory.

这意味着测试数据是通过使用 Timoshenko 梁理论中的真实杨氏模量生成的。

例句 2　This work specifically mines the data **generated by** humans while designing either a truss structure or an Internet-connected cooling system.

这项工作专门挖掘人类在设计桁架结构或互联网连接的冷却系统时产生的数据。

例句 3　Measurement values can be **generated from** these data to create a persona that can be used to assess the impact of a product on a consumer.

可以从这些数据中生成度量值，以创建一个可用于评估产品对消费者影响程度的角色(产品虚拟使用者)。

19. individual　[ˌɪndɪˈvɪdʒuəl]　adj. 单独的，个人的

例句 1　The **individual** devices were etched, rinsed and dried on a hot plate with care to avoid cracking.

每个器件都要在热板上小心地蚀刻、冲洗和干燥，以避免龟裂。

例句 2　Both behavioral studies were conducted in teams, but the current work treats data from these studies at the **individual** level.

这两项行为研究都是在团队中进行的，但目前的工作是在个人层面处理这些研究数据。

例句 3　In addition to linear system errors, there are also errors which are due to the limitations of the motors used to drive the **individual** axes.

除了线性系统误差外，还有一些误差是由用于驱动各个轴的电动机的局限性造成的。

20. initial　[ɪˈnɪʃl]　adj. 初始的

例句 1　Both groups were given an **initial design** for manufacturability training session.

两组均接受了可制造性培训课程的初步设计。

例句 2　The deviation is more evident during the retraction of the soft finger to its **initial position**, since the pneumatic supply is stopped and gravity becomes the dominant force acting on the finger.

软手指在回缩到其初始位置时，偏差更为明显，因为气动供应停止，重力成为

作用于手指上的主导力。

例句 3　The advantage of such models is their relationship to the physical process of material removal and their ability to predict the jet footprint whenever the **initial conditions** are known.

这类模型的优势在于它们与材料去除的物理过程间的关系，以及它们在已知初始条件的情况下预测喷射轨迹的能力。

21. input　［ˈɪnpʊt］　n. 输入

例句 1　The resulting tools rely on user **input** to provide process suggestions.

生成工具依靠用户输入来提供流程建议。

例句 2　These frequency sweep measurements are performed without any phase lock on the **input current**.

这些频率扫描测量是在输入电流没有任何锁相的情况下进行的。

例句 3　The **input signal** was chosen to have the complete hysteresis loop with minimal distortion due to the frequency.

选择的输入信号应具有完整的滞后环路，并且由频率所引起的失真最小。

22. linear　［ˈlɪniə(r)］　adj. 线性的

例句 1　The edge force is expressed as a **linear function** of the chip length.

边缘力被表达为芯片长度的线性函数。

例句 2　For this, the length of modified **linear segments** and inserted spline segment are first calculated.

为此，首先计算的是修改后的直线性段和插入的样条段的长度。

例句 3　However, if the overlapping is large, the **linear model** is not accurate enough to predict the profile.

然而，如果重叠度大，线性模型就无法足够准确地预测轮廓。

23. matrix　［ˈmeɪtrɪks］　n. 矩阵

例句 1　The codes are represented for each team as a **matrix of** values.

每组的代码都以数值矩阵的形式表示。

例句 2　By perturbing the spring constant of one resonator, two sets of eigenvalues can be measured to reconstruct the system **matrix of** the coupled system.

通过扰动一个谐振器的弹簧常数，可以测量两组特征值来重建耦合系统矩阵。

例句 3　The metal routing on each layer of the platform was replaced with a **matrix of** metal and oxide with the same effective percentage of metal contained in that part of the metal stack.

平台每层的金属走线被替换为金属和氧化物的矩阵，该部分金属堆中所包含的金属具有相同的有效百分比。

24. mechanism　［ˈmekənɪzəm］　n. 机制，机构

例句 1　Furthermore, a **mechanism for** enforcing kinematic constraints is incorporated into the proposed method.

此外，在提出的方法中引入了一种增强运动学约束的机制。

例句 2　Fig. 3 shows the **mechanism of** chip generation when the uncut chip approaches the honed edge of the cutting tool.

图 3 显示了未切割部分接近刀具珩磨刃时切屑的产生机理。

例句 3　Bennett addressed a four-bar **mechanism with** nonparallel and nonintersecting axes.

Bennett 提出了一种具有非平行轴和非相交轴的四杆机构。

25. mechanisms　［ˈmekənɪzəmz］　n. 机制（复数）

例句 1　In other words, these algorithms lend themselves well for designing complex **mechanisms** with constraints.

换句话说，这些算法很适合设计有约束的复杂机制。

例句 2　Evidence has been presented to show that the drilling **mechanisms of** the two drills are significantly different.

已有证据表明，这两种钻头的钻探机制有很大的不同。

例句 3　Depth profiling, digital microscopy and scanning electron microscopy are utilized to investigate topological evolution and **mechanisms of** grit failure.

深度剖析、数字显微镜和扫描电子显微镜被用于研究磨粒破坏的拓扑演化和机制。

26. methods　［ˈmeθədz］　n. 方法（复数）

例句 1　Future research can focus on better **methods to** compute similarities.

未来的研究可专注于更好的计算相似度的方法。

例句 2　Many involve **methods with** varying degrees of process planning or production rules.

许多方法涉及具有不同程度的工艺计划或生产规则。

例句 3　In practice, most widely used **methods of** computing similarity between vectors, such as cosine, radial basis function, or Hamming distances satisfy this condition.

在实践中，大多数广泛使用的计算向量之间相似性的方法，如余弦、径向基函数或汉明距离等，都满足这一条件。

27. minimum　['mɪnɪməm]　adj. 最小的，最低限度的

例句 1　**A minimum of** three linkages ensures proper bracing in all directions.

最起码要有3个连杆才可确保在各个方向提供足够支撑。

例句 2　Therefore we need **a minimum of** measurements to extract their values.

因此，我们需要用最低限度测量来提取它们的值。

例句 3　However, at low doping levels, the temperature at which the resistivity reaches a **minimum value**, is higher (around −200℃).

然而，在低掺杂水平下，电阻率达到最低值的温度更高(约-200℃)。

28. negative　['neɡətɪv]　adj. 负的，负面的

例句 1　This is due to the highly **negative rake angle** on the tip of the tool.

这是由于刀尖上的高度负前角造成的。

例句 2　However, the **negative impact** of AM on the design process is largely not discussed in literature.

然而，文献中有关增材制造对设计过程负面影响的内容很少。

例句 3　By identifying the dimensions with a **negative impact**, the product can be changed and improved to increase its positive impact.

通过确定具有负面影响的维度，可以对产品进行更改和改进，以增加其积极影响。

29. obtained　[əb'teɪnd]　v. 获得，得到(过去分词)

例句 1　Our result is similar to the result **obtained by** Sohn et al.

我们的结果与Sohn等人得到的结果相似。

例句 2　The results **obtained from** this method are listed in Table 7.

通过该方法得到的结果见表7。

例句 3   In this case, path error is calculated from individual axis errors and a control action is then **obtained for** the path error.

在这种情况下，路径误差是由单个的轴误差计算出来的，然后得到一个针对路径误差的控制动作。

30. parameter   [pəˈræmɪtə(r)]   n. 参数

例句 1   The physical **parameter values** are listed in Table A2.

表 A2 中列出了物理参数值。

例句 2   These **parameter values** were used for the demand simulation.

这些参数值被用于需求模拟。

例句 3   Therefore, in this paper, both prediction accuracy and **parameter estimation accuracy** are evaluated as two important performances for calibration methods.

因此，本文将预测精度和参数估计精度作为标定方法的两个重要性能进行评价。

31. previous   [ˈpriːviəs]   adj. 以前的，先前的

例句 1   This result agrees with the **previous work** carried out with polyethylene glycol.

这一结果与以前用聚乙二醇进行的研究一致。

例句 2   These samples were deposited using a doctor blade technique in order to compare them to **previous research**.

这些样品是用刮刀技术沉积的，以便与以前的研究进行比较。

例句 3   As discussed in the **previous section**, the print quality is best when the feed direction aligns with the scan velocity.

正如前一节所讨论的，当进给方向与扫描速度一致时，打印质量最好。

32. range   [reɪndʒ]   n. 范围；v. 从……到……

例句 1   The grain size of the drill material was measured and found to be **in the range of** 500-600 nm.

经测量，钻头材料的晶粒尺寸范围为 500~600 nm。

例句 2   These techniques **range from** quite simple tasks to incredibly complicated techniques.

这些技术包括从相当简单的任务到十分复杂的技术。

例句 3　Although specific frequencies did not match, the general **range of frequencies for all tests was consistent.**

虽然具体的频率并不匹配，但所有测试的总体频率范围是一致的。

33. ratio　['reɪʃiəʊ]　n. 比率

例句 1　The volume to surface area **ratio** is used as a measure of geometric complexity.

体积与表面积的比率被用以衡量几何复杂性。

例句 2　More than 20% improvement in sensitivity was demonstrated with a coupling **ratio of** 1.6.

在耦合比为 1.6 的情况下，灵敏度提高了 20% 以上。

例句 3　The **ratio of** steady-state error **to** the desired force is constant (in contrast to the open-loop controller).

稳态误差与期望力的比率是恒定的(与开环控制器相反)。

34. reference　['refrəns]　n. 参考

例句 1　**In reference to** Table 3, each edge is graded with a weight from 0.2 to 1.

参照表 3，将每条边的权重从 0.2 到 1 进行分级。

例句 2　The choice of the **reference point** used is important and can significantly affect the interpretations about relative performance.

所用的参考点的选择很重要，可以显著影响对相对性能的解释。

例句 3　Physically, this system delays the adding of **reference position** in all axes until all necessary conditions are satisfied.

在物理上，该系统延迟添加所有轴的参考位置，直到满足所有必要条件。

35. research　[rɪ'sɜːtʃ]　n. 研究

例句 1　The guiding **research questions** and conceptual framework influenced data collection, as shown in Fig. 1.

如图 1 所示，引导性研究问题和概念框架影响了数据收集。

例句 2　One line of **research in** the area is the development of novel mathematical techniques to build accurate and generalizable surrogate models.

该领域的一个研究方向是开发新的数学技术，以建立准确和可推广的代理模型。

例句 3 To date, the **research on** coupled resonant sensors has concentrated on weakly coupled systems, which rely on the measurement of signal amplitudes.

迄今为止,对耦合谐振传感器的研究主要集中在依赖信号振幅测量的弱耦合系统之上。

36. significantly [sɪgˈnɪfɪkəntli] adv. 明显地

例句 1 This arrangement is likely to offer a **significantly different** cutting action to that offered by diamond abrasives.

这种排列提供的切削作用有可能与金刚石磨料所提供的明显不同。

例句 2 Considering the magnitude of shrinkage for a single layer, the overall shrinkage for the build part would be **significantly higher**.

考虑到单层的收缩幅度,构建部件的整体收缩将显著提高。

例句 3 The driving voltage was **significantly lower** than those previously reported with similar device dimensions using single-sided electrode designs.

驱动电压明显低于以前报道的使用单面电极设计的类似装置尺寸的电压。

37. simulation [ˌsɪmjuˈleɪʃ(ə)n] n. 仿真,模拟(实验)

例句 1 The **simulation results** are plotted in Fig. 9.

仿真结果如图 9 所示。

例句 2 The **simulation results** show the similar trend.

仿真结果显示了类似的趋势。

例句 3 The specimens also have the same geometry of the **simulation model**.

试样也具有与仿真模型相同的几何形状。

38. source [sɔːs] n. 来源

例句 1 Consumer reviews are a potentially rich **source of** information.

消费者评论是个丰富的潜在信息来源。

例句 2 One **source of** disturbance could be from environments such as air flow.

扰动可能是源自环境,比如空气流动。

例句 3 Another **source of** difficulty comes from cultural differences between the engineer and consumers.

另一个难题来源于工程师和消费者之间的文化差异。

39. technique ［tekˈniːk］ n. 技术

例句1　Protocol analysis is a common **technique for** studying design activity empirically.

协议分析是实证研究设计活动的一种常用技术。

例句2　Further research should expand the reduction **technique to** greater generality.

进一步的研究应将简化技术推广至更广泛的应用范围。

例句3　To prevent such undesired change, this paper proposes a **technique to** couple an electrical resonator to an array of MEMS resonators.

为了防止这种不利改变的发生，本文提出了一种将电谐振器耦合到微电子机械系统谐振器阵列的技术。

40. techniques ［tekˈniːks］ n. 技术（复数）

例句1　Numerous optimization **techniques** have been developed with two major groups being studied intensively.

随着两大主要群体的深入研究，许多优化技术得到了发展。

例句2　The hypothesis is tested using a multi-agent simulation for data generation and machine learning **techniques for** data analysis.

用多代理模拟进行数据生成以及用机器学习技术进行数据分析来对假设进行验证。

例句3　Previously proposed **techniques to** characterize coupled system use either the inherent non-linearity of the system or physically change the system.

以前提出的表征耦合系统技术通过系统固有的非线性或物理方法改变系统。

41. torque ［tɔːk］ n. 扭矩

例句1　However, our target is to control the **output torque**.

然而，我们的目标是控制输出转矩。

例句2　In typical applications, the maximum **torque capacity** of the brake is used to stop the rotating system altogether.

在典型的应用中，制动器的最大扭矩能力用于停止整个旋转系统。

例句3　A second drilling dynamometer was placed below the vice in order to record thrust forces and **torque of** a drilling process.

在虎钳下方放置第二个钻井测力计，用以记录钻井过程中的推力和扭矩。

42. volume ['vɒljuːm] n. 体积，容量

例句1 In addition, the small **volume of** these devices results in low heat capacities.

此外，这些设备的小容积导致热容量低。

例句2 These systems use a small **volume of** reactive adhesive to glue the workpiece to a set of fixturing datum pins.

这些系统使用少量的反应性黏合剂将工件粘在一组固定基准销上。

例句3 The **volume fraction equation** was solved along with energy and momentum equation to predict temperature, fluid flow and surface deformation.

通过求解体积分数方程、能量和动量方程来预测温度、流体流动和表面变形。

第四部分 一般学术词汇

词汇表

单词	音标	词性	词义
acceleration	[əkˌseləˈreɪʃn]	n.	加速度
accelerometer	[əkˌseləˈrɒmɪtə]	n.	加速度计
accurate	[ˈækjərɪt]	adj.	准确的
accurately	[ˈækjərətli]	adv.	准确地
achieved	[əˈtʃiːvd]	v.	实现(过去分词)
adaptive	[əˈdæptɪv]	adj.	能适应的
additionally	[əˈdɪʃənəli]	adv.	此外
affect	[əˈfekt]	v.	影响
affected	[əˈfektɪd]	v.	影响(过去分词)
algorithms	[ˈælgərɪðəmz]	n.	算法(复数)
alternative	[ɔːlˈtɜːnətɪv]	adj. /n.	可供替代的，另一种的/替代物
analog	[ˈænəˌlɒg]	adj.	模拟的
analytical	[ˌænəˈlɪtɪkəl]	adj.	分析的
analyzed	[ˈænəlaɪzd]	v.	分析(过去分词)
angular	[ˈæŋgjələ]	adj.	有棱角的
appendix	[əˈpendɪks]	n.	附录

续表

单词	音标	词性	词义
approaches	[əˈprəʊtʃɪz]	n.	方法(复数)
appropriate	[əˈprəʊpriət]	adj.	适当的
approximately	[əˈprɒksɪmətli]	adv.	大约
architecture	[ˈɑːkɪˌtɛktʃə]	n.	结构
areas	[ˈeriəz]	n.	地区，区域(复数)
assumed	[əˈsuːmd]	v.	假设(过去分词)
assuming	[əˈsjuːmɪŋ]	conj.	假设……为真；假如
assumption	[əˈsʌmpʃn]	n.	假设
attached	[əˈtætʃt]	adj.	附着的
authors	[ˈɔːθəz]	n.	作者(复数)
available	[əˈveɪləbl]	adj.	可获得的
bandwidth	[ˈbændˌwɪdʒ]	n.	带宽
calibration	[ˌkælɪˈbreɪʃn]	n.	校准
capabilities	[ˌkeɪpəˈbɪlətiz]	n.	能力(复数)
capability	[ˌkeɪpəˈbɪlɪti]	n.	能力
capable	[ˈkeɪpəbl]	adj.	有能力的
carbide	[ˈkɑːbaɪd]	n.	碳化物
carrier	[ˈkærɪə]	n.	载体
cell	[sel]	n.	单元
cells	[selz]	n.	单元(复数)
challenges	[ˈtʃælɪndʒɪz]	n.	挑战(复数)
chemical	[ˈkɛmɪkəl]	adj.	化学的
coefficients	[ˌkəʊɪˈfɪʃənts]	n.	系数(复数)
compensation	[ˌkɒmpenˈseɪʃn]	n.	补偿

续表

单词	音标	词性	词义
complex	[ˈkɒmpleks]	adj.	复杂的
complexity	[kəmˈpleksɪti]	n.	复杂性(度)
compliance	[kəmˈplaɪəns]	n.	柔度
component	[kəmˈpəʊnənt]	n.	组成部分
computed	[kəmˈpjuːtɪd]	v.	计算(过去分词)
concept	[ˈkɒnsept]	n.	概念
conclusion	[kənˈkluːʒn]	n.	结论
conclusions	[kənˈkluːʒənz]	n.	结论(复数)
conducted	[kənˈdʌktɪd]	v.	实施(过去分词)
conductivity	[ˌkɑːndʌkˈtɪvəti]	n.	传导性
configuration	[kənˌfɪɡjʊˈreɪʃn]	n.	配置
consistent	[kənˈsɪstənt]	adj.	一致的
consists	[kənˈsɪsts]	v.	由……组成(单数第三人称一般现在时)
constraints	[kənˈstreɪnts]	n.	约束(复数)
constructed	[kənˈstrʌktɪd]	v.	构建(过去分词)
construction	[kənˈstrʌkʃn]	n.	建造；构建
contrast	[ˈkɑːntræst]	n.	对比
controllers	[kənˈtrəʊləz]	n.	控制器(复数)
create	[kriˈeɪt]	v.	创造
created	[kriˈeɪtɪd]	v.	创造(过去分词)
creating	[kriˈeɪtɪŋ]	v.	创造(动名词)
curvature	[ˈkɜːvətʃə]	n.	曲率
cycle	[ˈsaɪkəl]	n.	周期

续表

单词	音标	词性	词义
decomposition	[ˌdiːkɒmpəˈzɪʃn]	n.	分解
define	[dɪˈfaɪn]	v.	定义
deformation	[ˌdiːfɔːˈmeɪʃn]	n.	变形
demonstrate	[ˈdemənˌstreɪt]	v.	表明，演示
demonstrated	[ˈdemənstreɪtɪd]	v.	表明(过去分词)
dense	[dens]	adj.	稠密的
density	[ˈdensɪtɪ]	n.	密度
deposition	[ˌdepəˈzɪʃn]	n.	沉积
derived	[dɪˈraɪvd]	v.	导出，得出(过去分词)
designed	[dɪˈzaɪnd]	v.	设计(过去分词)
designs	[dɪˈzaɪnz]	n.	设计(复数)
detection	[dɪˈtekʃn]	n.	检测
deviation	[ˌdiːviˈeɪʃn]	n.	偏差
device	[dɪˈvaɪs]	n.	装置，设备
devices	[dɪˈvaɪsɪz]	n.	装置，设备(复数)
differential	[ˌdɪfəˈrenʃl]	adj.	微分的
digital	[ˈdɪdʒɪtl]	adj.	数字的，数码的
dimensional	[dɪˈmenʃənl]	adj.	尺寸的
dimensions	[daɪˈmenʃ(ə)nz]	n.	尺寸；维度(复数)
displacement	[dɪsˈpleɪsmənt]	n.	位移
distributed	[dɪˈstrɪbjuːtɪd]	v.	分布，分散(过去分词)
distribution	[ˌdɪstrɪˈbjuːʃn]	n.	分布
domain	[dəʊˈmeɪn]	n.	域
dynamic	[daɪˈnæmɪk]	adj.	动态的

续表

单词	音标	词性	词义
electron	[ɪˈlektrɒn]	n.	电子
element	[ˈelɪmənt]	n.	元素
embedded	[ɪmˈbedɪd]	v.	嵌入（过去分词）
empirical	[ɛmˈpɪrɪkəl]	adj.	实证的，以经验为依据的
enable	[ɪnˈeɪbəl]	v.	使能够
ensure	[ɛnˈʃʊə]	v.	确保
environment	[ɪnˈvaɪrənmənt]	n.	环境
environments	[ɪnˈvaɪrənmənts]	n.	环境（复数）
equivalent	[ɪˈkwɪvələnt]	adj.	相等的，相当的
established	[ɪˈstæblɪʃt]	v.	确立，建立（过去分词）
estimate	[ˈestɪmeɪt]	v.	估计
estimated	[ˈestɪmeɪtɪd]	v.	估计（过去分词）
estimation	[ˌɛstɪˈmeɪʃn]	n.	估计
etched	[etʃt]	v.	蚀刻（过去分词）
etching	[ˈetʃɪŋ]	v.	蚀刻（现在分词）
evaluate	[ɪˈvæljʊeɪt]	v.	评估，估计
evaluated	[ɪˈvæljʊeɪtɪd]	v.	评估，估计（过去分词）
evaluation	[ɪˌvæljuˈeɪʃn]	n.	评价，评估
evolution	[ˌiːvəˈluːʃn]	n.	演变
expansion	[ɪkˈspænʃn]	n.	扩张，膨胀
experimentally	[ɪkˌsperɪˈmentəli]	adv.	实验上地
external	[ɪkˈstɜːrnl]	adj.	外部的
extracted	[ɪkˈstræktɪd]	v.	提取（过去分词）
fabricated	[ˈfæbrɪkeɪtɪd]	v.	制造（过去分词）

续表

单词	音标	词性	词义
fabrication	[ˌfæbrɪˈkeɪʃn]	n.	制造
factors	[ˈfæktəz]	n.	因素(复数)
feasibility	[ˌfiːzəˈbɪləti]	n.	可行性
feasible	[ˈfiːzəbl]	adj.	可行的
feature	[ˈfiːtʃə(r)]	n.	特征
features	[ˈfiːtʃə(r)z]	n.	特征(复数)
feedback	[ˈfiːdbæk]	n.	反馈
finally	[ˈfaɪnəli]	adv.	最后
finite	[ˈfaɪnaɪt]	adj.	有限的，有限制的
flexibility	[ˌfleksəˈbɪləti]	n.	灵活性
focus	[ˈfoʊkəs]	v.	集中，关注；聚焦
focused	[ˈfoʊkəst]	v.	集中，关注；聚焦（过去分词）
framework	[ˈfreɪmwɜːrk]	n.	结构，框架
frequencies	[ˈfrikwənsiz]	n.	频率(复数)
functions	[ˈfʌŋkʃnz]	n.	功能、函数(复数)
furthermore	[ˌfɜːrðərˈmɔː(r)]	adv.	此外，而且
generate	[ˈdʒenəreɪt]	v.	产生，引起
generation	[ˌdʒenəˈreɪʃn]	n.	产生；(产品发展的)代
geometric	[ˌdʒiːəˈmetrɪk]	adj.	几何的
geometry	[dʒiˈɑːmətri]	n.	几何；几何形状
global	[ˈgloʊbl]	adj.	全局的
height	[haɪt]	n.	高；高度
hence	[hens]	adv.	因此；之后

续表

单词	音标	词性	词义
identical	[aɪˈdentɪkl]	adj.	完全相同的
identification	[aɪˌdentɪfɪˈkeɪʃn]	n.	辨认，识别
identified	[aɪˈdentɪfaɪd]	v.	认出，识别（过去分词）
identify	[aɪˈdentɪfaɪ]	v.	认出，识别
illustrated	[ˈɪləstreɪtɪd]	v.	说明；表明；显示（过去分词）
images	[ˈɪmɪdʒɪz]	n.	图像；影像；肖像（复数）
impact	[ˈɪmpækt]	n.	影响
implementation	[ˌɪmplɪmenˈteɪʃn]	n.	实施，执行
implemented	[ˈɪmplɪmentɪd]	v.	实施（过去分词）
index	[ˈɪndeks]	n.	指数
indicate	[ˈɪndɪkeɪt]	v.	表明，指示
indicates	[ˈɪndɪkeɪts]	v.	表明，指示（第三人称单数一般现在时）
indicating	[ˈɪndɪkeɪtɪŋ]	v.	表明，指示（现在分词）
induced	[ɪnˈdjʊst]	v.	诱发（过去分词）
instance	[ˈɪnstəns]	n.	例子，实例
integration	[ˌɪntɪˈɡreɪʃn]	n.	结合，整合
interaction	[ˌɪntərˈækʃn]	n.	互动，交流；相互影响，相互作用
internal	[ɪnˈtɜːrnl]	adj.	内部的
inverse	[ˌɪnˈvɜːrs]	adj.	相反的
investigate	[ɪnˈvestɪɡeɪt]	v.	调查研究
investigated	[ɪnˈvestɪɡeɪtɪd]	v.	研究（过去分词）
issue	[ˈɪʃuː]	n.	问题

续表

单词	音标	词性	词义
issues	[ˈɪʃuːz]	n.	问题(复数)
lateral	[ˈlætərəl]	adj.	横向的；侧面的
layer	[ˈleɪə(r)]	n.	层；层次
layers	[ˈleɪə(r)z]	n.	层；层次(复数)
located	[ˈloʊkeɪtɪd]	v.	位于（过去分词）
location	[loʊˈkeɪʃn]	n.	位置
machined	[məˈʃiːnd]	v.	(用机器)加工(过去分词)
machining	[məˈʃiːnɪŋ]	n.	(用机器)加工
magnitude	[ˈmæɡnɪtuːd]	n.	规模，大小
maintain	[meɪnˈteɪn]	v.	保持，维持
manual	[ˈmænjuəl]	adj.	手动的；人工的
methodology	[ˌmeθəˈdɒlədʒi]	n.	方法
micro	[ˈmaɪkroʊ]	adj.	微小的
microscopy	[maɪˈkrɒskəpi]	n.	显微镜
minimize	[ˈmɪnɪmaɪz]	v.	使最小化
modes	[moʊdz]	n.	模式(复数)
modified	[ˈmɑːdɪfaɪd]	v.	修改（过去分词）
modules	[ˈmɔdʒulz]	n.	模块(复数)
monitoring	[ˈmɒnətərɪŋ]	n.	监测
morphology	[mɔːrˈfɑːlədʒi]	n.	形态学
mounted	[ˈmaʊntɪd]	v.	安装（过去分词）
network	[ˈnetwɜːrk]	n.	网络
nominal	[ˈnɒmɪnl]	adj.	标称的
nonlinear	[ˌnɒnˈlɪniər]	adj.	非线性的

续表

单词	音标	词性	词义
normal	[ˈnɔːrml]	adj.	普通的；垂直的
normalized	[ˈnɔːrməˌlaɪzd]	v.	标准化；归一化（过去分词）
novel	[ˈnɒvl]	adj.	新颖的
objective	[əbˈdʒektɪv]	adj./n.	客观的；目的，目标
obtain	[əbˈteɪn]	v.	得到
occur	[əˈkɜːr]	v.	发生
occurs	[əˈkɜːrz]	vi.	发生，出现（第三人称单数一般现在时）
offset	[ˈɔːfset]	n.	偏移；抵消
optical	[ˈɑːptɪkl]	adj.	光学的
optimal	[ˈɒptɪməl]	adj.	最佳的
optimization	[ˌɒptɪməˈzeɪʃn]	n.	最优化
optimized	[ˈɒptɪmaɪzd]	v.	使最优化（过去分词）
orientation	[ˌɔːriənˈteɪʃn]	n.	方向
overall	[ˌoʊvərˈɔːl]	adj.	总的，整体的
overlap	[ˌoʊvərˈlæp]	n.	重叠部分
parallel	[ˈpærəlel]	adj.	平行的
peak	[piːk]	n.	峰值
physical	[ˈfɪzɪkl]	adj.	物理的
plastic	[ˈplæstɪk]	adj.	塑性的
plot	[plɒt]	n.	（表现两个变量关系的）图表
plotted	[plɒtɪd]	v.	标绘（过去分词）
polymer	[ˈpɑːlɪmər]	n.	[高分子]聚合物
porosity	[pɔːˈrɑːsəti]	n.	有孔性

续表

单词	音标	词性	词义
positive	[ˈpɑːzətɪv]	adj.	正向的
potential	[pəˈtenʃl]	n./adj.	潜力/潜在的
potentially	[pəˈtenʃəli]	adv.	潜在地
precision	[prɪˈsɪʒn]	n.	精度
predict	[prɪˈdɪkt]	v.	预测
predicted	[prɪˈdɪktɪd]	v.	预测（过去分词）
prediction	[prɪˈdɪkʃn]	n.	预测
predictions	[prɪˈdɪkʃənz]	n.	预测（复数）
previously	[ˈpriːviəsli]	adv.	先前地
primary	[ˈpraɪməri]	adj.	主要的
prior	[ˈpraɪə(r)]	adj.	先前的，在前面的
procedure	[prəˈsiːdʒə(r)]	n.	程序
processes	[ˈprɒsesɪz]	n.	过程，工艺（复数）
processing	[ˈprəʊsesɪŋ]	n.	加工，处理
profile	[ˈprəʊfaɪl]	n.	轮廓，剖面
proportional	[prəˈpɔːʃnl]	adj.	成比例的
radial	[ˈreɪdiəl]	adj.	径向的
real-time	[ˌriːəl ˈtaɪm]	adj.	实时的
region	[ˈriːdʒən]	n.	区域
regions	[ˈriːdʒənz]	n.	区域（复数）
removal	[rɪˈmuːvl]	n.	去除
remove	[rɪˈmuːv]	v.	去除
removed	[rɪˈmuːvd]	v.	去除（过去分词）
require	[rɪˈkwaɪə(r)]	v.	需要
requirements	[rɪˈkwaɪəmənts]	n.	需求（复数）

续表

单词	音标	词性	词义
requires	[rɪˈkwaɪəz]	v.	需要（第三人称单数一般现在时）
researchers	[ˈrɪsɜːtʃəz]	n.	研究人员（复数）
resolution	[ˌrezəˈluːʃn]	n.	清晰度，分辨率
response	[rɪˈspɒns]	n.	响应
responses	[rɪˈspɒnsɪz]	n.	响应（复数）
rigid	[ˈrɪdʒɪd]	adj.	刚性的
robust	[rəʊˈbʌst]	adj.	稳健的，坚固的
role	[rəʊl]	n.	作用
rotation	[rəʊˈteɪʃn]	n.	旋转
rotational	[rəʊˈteɪʃnl]	adj.	旋转的
scan	[skæn]	n.	扫描
scanning	[ˈskænɪŋ]	adj.	扫描的
sections	[ˈsekʃnz]	n.	部分（复数）
segments	[ˈsegmənts]	n.	片段；段数（复数）
selected	[sɪˈlektɪd]	v.	选择（过去分词）
selection	[sɪˈlekʃn]	n.	选择
sensing	[ˈsensɪŋ]	n.	传感，感应
sensors	[ˈsensəz]	n.	传感器（复数）
series	[ˈsɪəriːz]	n.	系列
setup	[ˈsetʌp]	n.	装置，安装，建立
similarly	[ˈsɪmələli]	adv.	相似地，同样地，也
simulated	[ˈsɪmjuleɪtɪd]	v.	模拟，模仿（过去分词）
simulations	[ˌsɪmjuˈleɪʃnz]	n.	模拟，仿真（复数）
simultaneously	[ˌsɪm(ə)lˈteɪniəsli]	adv.	同时地

续表

单词	音标	词性	词义
software	[ˈsɒftweə(r)]	n.	软件
specific	[spəˈsɪfɪk]	adj.	具体的
specifically	[spəˈsɪfɪkli]	adv.	具体地，特别地
specified	[ˈspesɪfaɪd]	v.	特定，指定（过去分词）
stability	[stəˈbɪləti]	n.	稳定性
stable	[ˈsteɪbl]	adj.	稳定的
static	[ˈstætɪk]	adj.	静态的
strain	[streɪn]	n.	应变，形变
strategy	[ˈstrætədʒi]	n.	策略
stress	[stres]	n.	应力
structural	[ˈstrʌktʃərəl]	adj.	结构的
structures	[stˈrʌktʃəz]	n.	结构（复数）
substrate	[ˈsʌbstreɪt]	n.	基板，基底
sufficient	[səˈfɪʃnt]	adj.	充分的，足够的
summarized	[ˈsʌməraɪzd]	v.	总结，概括（过去分词）
tangential	[tænˈdʒenʃl]	adj.	切线的，切向的
target	[ˈtɑːgɪt]	n.	目标
task	[tɑːsk]	n.	任务
tasks	[tɑːsks]	n.	任务（复数）
technologies	[tekˈnɒlədʒiz]	n.	技术（复数）
technology	[tekˈnɒlədʒi]	n.	技术
theoretical	[ˌθɪəˈretɪkl]	adj.	理论的
theory	[ˈθɪəri]	n.	理论
thermal	[ˈθɜːm(ə)l]	adj.	热的，热量的
threshold	[ˈθreʃhəʊld]	n.	阈值

续表

单词	音标	词性	词义
tissue	[ˈtɪʃuː]	n.	组织
traditional	[trəˈdɪʃnl]	adj.	传统的
transfer	[trænsˈfɜː(r)]	n.	传递，传导
transformation	[ˌtrænsfəˈmeɪʃn]	n.	变化，变换
transient	[ˈtrænziənt]	n.	瞬态
transition	[trænˈzɪʃn]	n.	转变，转换
trend	[trend]	n.	趋势
uniform	[ˈjuːnɪfɔːm]	adj.	均匀的
utilized	[ˈjuːtəlaɪzd]	v.	采用，利用（过去分词）
vacuum	[ˈvækjuːm]	adj./n.	真空的；真空
validate	[ˈvælɪdeɪt]	v.	验证，证实
validated	[ˈvælɪdeɪtɪd]	v.	验证，证实（过去分词）
validation	[ˌvælɪˈdeɪʃn]	n.	验证
variable	[ˈveəriəbl]	adj./n.	不同的，可变的；变量
variation	[ˌveəriˈeɪʃn]	n.	变化
variations	[ˌveəriːˈeɪʃnz]	n.	变化（复数）
vary	[ˈveəri]	v.	（使）不同，变化
varying	[ˈveəriɪŋ]	v.	变化（现在分词或动名词）
vectors	[ˈvɛktəz]	n.	矢量，向量（复数）
velocities	[vəˈlɑsətiz]	n.	速度（复数）
versus	[ˈvɜːsəs]	prep.	与……相比
vertical	[ˈvɜːtɪkl]	adj.	垂直的
via	[ˈvaɪə]	prep.	通过，经由
vibration	[vaɪˈbreɪʃn]	n.	振动
whereas	[ˌweərˈæz]	conj.	然而

词汇详解

1. acceleration [ækˌsɛləˈreɪʃn] n. 加速度

例句1　Kinematics of the axes during corner transition is dictated by the jerk limited **acceleration profile**.

拐角过渡期间，轴的运动由加速度限制曲线决定。

例句2　Fig. 10 illustrates the frequencies and the relative amplitudes of the sinusoids present in each axis **acceleration profile**.

图10阐释了每个轴加速度曲线中存在的正弦波的频率和相对振幅。

2. accelerometer [ækˌsɛləˈrɒmɪtə] n. 加速度计

例句1　Considering the small **accelerometer**, the c values increase from 0.07 to 0.17 to 0.85, respectively, for these three setups.

考虑到加速度计较为微小，对于这3种设置，c 值从 0.07 分别增加到 0.17 再到 0.85。

例句2　The damping factor of the device was tailored using dedicated damping electrodes to ensure stable operation of the quasi-static **accelerometer** at low-pressure.

该装置的阻尼系数是使用专用阻尼电极定制的，以确保准静态加速度计在低压下的稳定运行。

3. accurate [ˈækjuərɪt] adj. 准确的

例句1　As expected, the figure shows that with more data, more **accurate predictions** are possible.

正如预期的那样，该图显示：随着数据增加，有可能做出更准确的预测。

例句2　This in turn allowed them to better estimate the final quality of early concepts, providing more **accurate information** to guide the search process.

这反过来又使他们能够更好地估计早期概念的最终质量，为指导搜索过程提供更准确的信息。

4. accurately [ˈækjuərətli] adv. 准确地

例句1　However, it is important to **accurately identify** the model parameters for many reasons.

然而，出于多种原因，准确识别模型参数是很重要的。

例句2 See Eren [35] for more details on the number and variety of cutting process conditions needed to **accurately estimate** force model coefficients.

关于准确估计力模型系数所需切削工艺条件的数量和种类的更多细节，参阅 Eren [35]。

5. achieved [əˈtʃiːvd] v. 实现（过去分词）

例句1 This was **achieved by** translating the outer tube downward over the inner tube.

这是通过将外管向下平移到内管上来实现的。

例句2 Flexibility **is achieved through** the interchangeability of different modules, which enables a modular fixture to be easily disassembled and reassembled.

灵活性是通过不同模块的互换性实现的，这使模块化夹具可以很容易地被拆卸和重组。

6. adaptive [əˈdæptɪv] adj. 能适应的

例句1 At that moment, **adaptive control** has larger fluctuation because the parameters had not converged to their real values yet.

那时候，由于参数尚未收敛到实际值，自适应控制的波动较大。

例句2 This means that this **adaptive controller** does account for the tracking error, its changing speed, and its accumulation.

这意味着该自适应控制器确实导致跟踪误差、速度变化及其累积。

7. additionally [əˈdɪʃənəli] adv. 此外

例句1 **Additionally**, the electrical connections between the front and the handle side of the wafer were not available as a part of the process.

此外，晶圆片前端与手柄端之间的电气连接并不是该工艺的一部分。

例句2 **Additionally**, it was noted that while there was a high amount of gesturing in both systems, there were significant qualitative differences.

此外，值得注意的是，虽然两个系统中都有大量的手势，但在质量上却有很大差别。

8. affect [əˈfɛkt] v. 影响

例句1　However two error sources appear which **affect the accuracy** of the test.
然而，出现了两个影响测试准确性的误差源。

例句2　Artificial intelligence approaches require significant sets of data for training and give little insight into the details of the physics that **affect the process**.
人工智能方法需要大量的数据来进行训练，而且几乎无法让人深入了解影响流程的物理细节。

9. affected　［əˈfɛktɪd］　v. 影响(过去分词)

例句1　Thus, machined surface characteristics are greatly **affected by** the chip formation mechanism.
因此，加工的表面特征在很大程度上受到切屑形成机制的影响。

例句2　The nano-hardness was not significantly **affected by** an increase in cutting speeds and feed rate under dry and wet cutting conditions.
在干式和湿式切削条件下，提高切削速度和进给速度对纳米硬度的影响不大。

10. algorithms　［ˈælgəˌrɪðəmz］　n. 算法(复数)

例句1　Section 3 describes the clustering **algorithms** and evaluation approaches.
第3节描述了聚类算法和评估方法。

例句2　Control experiments of other closed-loop control **algorithms** have been also conducted to compare their performance with the new controller effectiveness.
还进行了其他闭环控制算法的控制实验，并与新控制器的有效性进行了性能比较。

11. alternative　［ɔːlˈtɜːnətɪv］　adj. 可供替代的，另一种的；n. 替代物

例句1　An **alternative approach** for reducing contour errors under such restrictive conditions is pre-compensation.
在这种限制性条件下，减少轮廓误差的另一种方法是预补偿。

例句2　It can also be considered as a viable **alternative** to conventional shear-based processes in appropriate circumstances.
在适当的条件下，它也可以被认为是传统剪切工艺的可行替代方案。

12. analog　［ˈænəˌlɒg］　adj. 模拟的

例句1　A digital **analog converter** was used for generating the scaled-down magnitude of the computed voltage.

采用数模转换器生成按比例缩小的计算电压。

例句2　Accelerometers need additional **analog to digital converters**（ADC），adding cost and complexity to the computer numerical control systems.

加速度计需要额外的模数转换器，这增加了计算机数字控制系统的成本和复杂性。

13. analytical　［ˌænəˈlɪtɪkəl］　adj. 分析的

例句1　However, under certain conditions a closed-form **analytical solution** can be achieved.

然而，在一定条件下，可以得到一个封闭形式的解析解。

例句2　This difference can be attributed to how the lumped circuit parameters of the **analytical model** are measured and modeled.

这种差异可以归因于分析模型的集总电路参数是如何测量和建模的。

14. analyzed　［ˈænəlaɪzd］　v. 分析（过去分词）

例句1　These two groups of tracks are **analyzed** separately.

分别对这两组轨道进行了分析。

例句2　A total of 185 frames from the camera recording are **analyzed with** this method to calculate the melt pool width.

利用该方法对摄像机记录的总计185帧的图像进行分析，计算出熔池宽度。

15. angular　［ˈæŋɡjʊlə］　adj. 有棱角的

例句1　The goal is to optimize this equation with respect to **angular position**.

我们的目标是优化这个关于角度位置的方程式。

例句2　Therefore, a planar linkage with **angular outputs** is able to provide such motions.

因此，具有角输出的平面连杆机构能够提供这种运动。

16. appendix　［əˈpɛndɪks］　n. 附录

例句1　The handout provided to the students is provided **in the appendix**.

附录中列出了给学生的讲义。

例句2　It is included **in appendix A** for not making the main body of this paper lengthy.

它被包含在附录 A 中,以避免本文的正文部分冗长。

17. approaches [əˈprəʊtʃɪz] n. 方法(复数)

例句 1　The following sections outline the **approaches to** minimise the set-up errors in the spindle tool cup and the centre pivot tool cup.

以下部分概述了将主轴工具杯和中心枢轴工具杯的安装误差降到最低的方法。

例句 2　There is another set of more thorough **approaches for** predicting footprint profiles based on analytical geometric modelling.

还有另一套更全面的方法通过分析几何建模来预测印迹轮廓。

18. appropriate [əˈprəʊpriət] adj. 适当的

例句 1　By an **appropriate** choice of the modes to be included in the target surface, one can ensure that there is a solution of the inverse problem.

通过对目标面所包含的模式进行适当选择,可以确保有逆问题的解决方案。

例句 2　The bubble eventually reached a size that was **appropriate for** discharging and became a stable gas film supporting the electrochemical discharge phenomenon.

气泡最终达到了适合放电的大小,成为支持电化学放电现象的稳定气膜。

19. approximately [əˈprɒksɪmətli] adv. 大约

例句 1　On average the grains extend at **approximately** 601 nm from the radial direction.

平均而言,晶粒从径向方向延伸约 601 nm。

例句 2　It also assumes that the local radius of curvature at the reference and actual position are **approximately equal**.

它还假设参考位置和实际位置的局部曲率半径大致相等。

20. architecture [ˈɑːkɪˌtɛktʃə] n. 结构

例句 1　Structural complexity of a system is closely related to the complexity of individual system elements and degree of connectivity of the underlying **system architecture**.

一个系统的结构复杂性与单个系统元素的复杂性以及基础系统架构的连接程度密切相关。

例句 2　It is based around a 3 channels **control architecture**, which transmits positions information from master to slave, and returns both slave position and environmental force.

它基于一个三通道控制架构，主机向从机传输位置信息，并将从机位置和环境力传回。

21. areas ['eriəz] n. 地区，区域（复数）

例句 1 The **scan areas** of 1.2 mm were selected from different regions of the surface for the measurement.

从表面的不同部分选择 1.2 mm 的扫描区域进行测量。

例句 2 The graphical reachability results are generated by color coding the model **surface areas** based on their depth.

图形化的可达性结果是根据用颜色编码来表示不同深度的模型表面区域这一方式而实现的。

22. assumed [ə'suːmd] v. 假设（过去分词）

例句 1 Thus, the resultant cutting directions were **assumed to** be the same as the cutting velocity.

因此，合成的切削方向被假定与切削速度方向相同。

例句 2 **It is assumed that** this resistance is at the same temperature as the sense resistor and therefore does not contribute to an ambient temperature dependent error.

假设这个电阻与检测电阻的温度相同，因此它不会造成与环境温度有关的误差。

23. assuming [ə'sjuːmɪŋ] conj. 假设……为真；假如

例句 1 **Assuming that** the profile and kinematics of the grinding wheel are known, Li et al. modelled the arbitrary geometry of drill lips.

Li 等人假定砂轮的轮廓和运动学是已知的，对钻刃的任意几何形状进行了建模。

例句 2 **Assuming that** the system has proportional damping, following modal transformation decouples the equation of motion into its individual vibration modes.

假设系统具有比例阻尼，经过模态变换后，将运动方程解耦为其各个振动模态。

24. assumption [ə'sʌmpʃn] n. 假设

例句 1 The analysis is **based on the assumption that** these encoded relationships are correct.

分析的基础是假设这些编码的关系是正确的。

例句2　The measurement is **based on an assumption of** continuity in the temperature at the tool-workpiece interface and on the machined surface.

测量的基础是假设刀具-工件界面和加工表面的温度是连续的。

25. attached　[əˈtætʃt]　adj. 附着的

例句1　For simplicity, we refer to these particles as **attached** particles.

为了简单起见,我们把这些粒子称为附着粒子。

例句2　For these measurements, custom-made probe tips **attached to** micropositioners were used.

在这些测量中使用了附在微定位器上的定制探针尖。

26. authors　[ˈɔːθəz]　n. 作者(复数)

例句1　The **authors** demonstrate the relative efficiency of each approach experimentally using both corner tracking and circular interpolation as examples.

作者以角跟踪和圆弧插值为例,通过实验证明了每种方法的相对效率。

例句2　Additionally, the **authors** also highlight that these monitoring technologies can be used to determine tool wear and the stability of the manufacturing process.

此外,作者还强调这些监测技术可用于确定刀具磨损和加工过程的稳定性。

27. available　[əˈveɪləbl]　adj. 可获得的

例句1　Fuzzy logic reasoning has been used to monitor tool wear when multiple sensors are **available**.

在有多个传感器的情况下,模糊逻辑推理被用于监测刀具磨损。

例句2　This can all be determined without reading any reviews or even having the products **available for** physical measurement.

这一切都可以在不阅读任何评论或者甚至在没有产品可供实物测量的情况下确定下来。

28. bandwidth　[ˈbændˌwɪdz]　n. 带宽

例句1　However, the self-weight that each stage applies to the previous stage induces cross-coupling errors and limits operational **bandwidth**.

然而，每个阶段应用于前一阶段的自重会引起交叉耦合误差并限制操作带宽。

例句 2　Parylene based cantilevers have previously been developed for creating wide **bandwidth** electromagnetic energy harvesters.

基于聚对二甲苯的悬臂梁已被开发并用于制造宽带电磁能量收集器。

29. calibration　[ˌkælɪˈbreɪʃn]　n. 校准

例句 1　In this paper, we propose a sensitivity-based **calibration method**, which is simple enough for industrial practitioners.

本文中，我们提出了一种基于灵敏度的、对于工业从业者而言非常简单的校准方法。

例句 2　In the following discussions, two types of variables will be used, namely, design variable and **calibration parameter**.

下面的讨论中将使用两种类型的变量，即设计变量和校准参数。

30. capabilities　[ˌkeɪpəˈbɪlətiz]　n. 能力（复数）

例句 1　The result is buildable considering the current **capabilities of** AM processes.

考虑到当前增材制造工艺的先进程度，结果是可信的。

例句 2　An early academic report in relation to the integration of sensing **capabilities** into a work holding device was made in 1988 by Gupta et al.

早在 1988 年，Gupta 等人就发表了有关将传感能力集成到工件夹持装置的学术报告。

31. capability　[ˌkeɪpəˈbɪlɪtɪ]　n. 能力

例句 1　Both are especially promising since they have the **capability to** achieve a faster response.

这两种方法都极具前景，因为它们有能力实现更快的响应。

例句 2　However, neither possesses the **capability to** handle performance functions with strong variate interactions.

然而，两者都不具备处理具有强变量交互性能函数的能力。

32. capable　[ˈkeɪpəbl]　adj. 有能力的

例句 1　A novel model that is **capable to** simulate flow characteristics and rheological properties of high pressure coolant is proposed.

提出了一个能够模拟高压冷却剂的流动特性和流变特性的新模型。

例句2　This means that this statistical control method is **capable of** detecting a bad (chipped) tool edge very close to the time that it actually happens.

这意味着这种统计控制方法能够在非常接近实际发生的时间内检测出不良的（破损的）刀刃。

33. carbide　[ˈkɑːbaɪd]　n. 碳化物

例句1　The tool wear for wet drilling starts from the abrasion of coating from the tungsten **carbide** substrate.

湿法钻孔刀具的磨损始于碳化钨基底涂层的磨损。

例句2　This section reviews the thermoresistive effect in common sensing materials such as metals, silicon and **silicon carbide**.

本节将回顾常见传感材料（如：金属、硅和碳化硅）的热阻效应。

34. carrier　[ˈkærɪə]　n. 载体

例句1　However, at a higher temperature range (metal-like region), the free **carrier concentration** is saturated while the carrier mobility still decreases.

然而，在更高的温度范围内（类似金属的区域），自由载流子浓度达到饱和，而载流子迁移率仍然下降。

例句2　Both the piezoresistive effect in n-type and p-type semiconductors was investigated in which the **carrier concentration** varied from low doping to high doping levels.

研究了 n 型和 p 型半导体的压阻效应，其中载流子浓度在低掺杂到高掺杂水平之间变化。

35. cell　[sel]　n. 单元

例句1　Subtask allocation in the **hybrid cell** for collaborative assembly divides the entire assembly task into several subtasks.

协同装配混合单元的子任务分配将整个装配任务划分为若干个子任务。

例句2　To create the **unit cell** for each function, we begin with a mathematical rendering program to create a two-dimensional surface profile.

为了创建每个函数的单元格，我们首先使用一个数学绘制程序来创建一个二维表面轮廓。

36. cells ［selz］ n. 单元(复数)

例句1　Any type of suitable robots or multi-robots can be used in the **hybrid cells**.
任何类型的合适机器人或多机器人都可以用于混合单元中。

例句2　We focus our attention on examining triply periodic structures as the **unit cells** within the wider scaffold structure.
我们将注意力集中在检查作为更广泛支架结构中单元格的三周期结构。

37. challenges ［ˈtʃælɪndʒɪz］ n. 挑战(复数)

例句1　The **challenges** of single component design are categorized under the part-design category.
单组件设计的挑战被归入零件设计的类别中。

例句2　Yet these studies focus their reviews on the three **main challenges** individually rather than collectively.
然而，这些研究回顾聚焦于这三个主要挑战的个体而不是集体之上。

38. chemical ［ˈkɛmɪkəl］ adj. 化学的

例句1　The differing **chemical and physical properties of** the matrix and reinforcement can also affect processing and care must be taken to avoid unwanted chemical reactions.
不同的化学和物理性质的基质和增强剂也会影响加工，必须注意避免不必要的化学反应。

例句2　Generally, the requirements include good wear resistance, high hot hardness, good thermal shock properties and adequate **chemical stability** at elevated temperatures.
一般来说，这些要求包括良好的耐磨性、高的热硬度、良好的热冲击性能以及高温下足够的化学稳定性。

39. coefficients ［ˌkəʊɪˈfɪʃnts］ n. 系数(复数)

例句1　The **cutting force coefficients** are identified from non-symmetric drilling tests.
切削力系数是由非对称钻削试验确定的。

例句2　This paper builds upon our previous work by Cui et al. investigating the effects of tool wear on **force model coefficients**.

本文建立在之前由 Cui 等人有关刀具磨损对力模型系数影响的研究基础之上。

40. compensation ［ˌkɒmpɛnˈseɪʃn］ n. 补偿

例句1　In simulations, the traditional **pre-compensation** provides a 46% reduction in maximum.

在模拟中，传统的预补偿能减少的最大量是46%。

例句2　In this paper an approach for the **compensation of** the dynamics for a step in the profile is presented.

本文提出了一种对剖面中的一个步骤进行动态补偿的方法。

41. complex ［ˈkɒmplɛks］ adj. 复杂的

例句1　The **complex parts** impossible to fabricate with any conventional method can be fabricated.

可以制造出任何传统方法都不可能制造的复杂零件。

例句2　Both authentic and pseudo phase-change fixturing methods are highly attractive for thin workpieces with **complex geometries**.

真实和伪相变的装夹方法对于具有复杂几何形状的薄型工件非常有吸引力。

42. complexity ［kəmˈplɛksɪtɪ］ n. 复杂性（度）

例句1　Integrative **complexity** is defined as the difference between the total system complexity and the sum of complexities of individual modules.

综合复杂度被定义为总系统复杂度与单个模块复杂度之和这两者之间的差异。

例句2　The internal **complexity** is closely related to overall system design and is further divided into structural complexity, dynamic complexity, and organizational complexity.

内部复杂性与整体系统设计密切相关，并可进一步分为结构复杂性、动态复杂性和组织复杂性。

43. compliance ［kəmˈplaɪəns］ n. 柔度

例句1　Next, the **acoustic compliance** of the diaphragm is calculated.

接下来，计算膜片的声学柔度。

例句2　The latter can be the **compliance function** or the displacement of the structure at selected points.

后者可以是柔量函数或者是在选定点处结构的位移。

44. component ［kəmˈpəʊnənt］ n. 组成部分

例句 1 A global surrogate model of the performance function can then be built based on the summation of the approximated **component functions**.
然后可以根据近似分量函数的总和建立一个性能函数的全局代理模型。

例句 2 The system technologies are realized through the integration and support of hardware and software, which are referred to as **component technologies**.
本系统的技术是通过软、硬件的集成和支持来实现的，被称为组件技术。

45. computed ［kəmˈpjuːtɪd］ v. 计算（过去分词）

例句 1 Solution quality was automatically **computed** and displayed to participants.
溶液质量被自动计算并显示给参与者。

例句 2 The total displacement traveled can be **computed by** integrating the acceleration profile.
移动的总位移可以通过整合加速度剖面来计算。

46. concept ［ˈkɒnsɛpt］ n. 概念

例句 1 To the best of our knowledge, it is the first time that the **concept of** information reuse is extended to the registration step.
据我们所知，这是第一次将信息重用的概念扩展到注册步骤。

例句 2 Based on the **concept of** defined cutting edges in solid diamond, there is the possibility to make significant improvements in core drilling performance.
基于实心金刚石中边界清晰的切削刃的概念，岩心钻探的性能有可能得以显著改善。

47. conclusion ［kənˈkluːʒn］ n. 结论

例句 1 **In conclusion**, we assert that additive manufacturing has a great deal to offer engineers and mechanical designers.
总之，我们断言，增材制造对工程师和机械设计师有很大的帮助。

例句 2 The **main conclusion** that can be drawn from this analysis is that melt pool width tends to decrease from the beginning of each track towards the end of each track.
从分析中可以得出的主要结论是熔融池的宽度从每条熔道的起点到终点呈减小

的趋势。

48. conclusions ［kənˈkluːʒənz］ n. 结论（复数）

例句1 **The following conclusions can be made on** tool wear for the micro-drilling process for nickel alloys.

对于镍合金微钻加工的刀具磨损问题，可以得出以下结论。

例句2 Section 4 applies the proposed method to engineering application on crush simulation of honeycomb tube, **followed by conclusions in** Sec. 5.

第4节将所提出的方法用于蜂窝管挤压模拟的工程应用中，随后在第5节中给出结论。

49. conducted ［kənˈdʌktɪd］ v. 实施（过去分词）

例句1 Bracing tests were **conducted by** applying force to the catheter inside the brace and measuring the resulting displacement.

通过对支架内的导管施力并测量所产生的位移来实施支架测试。

例句2 Experiments were **conducted on** an ultra-precision lathe machine with superior rigidity and stability in both translational and rotational motions.

实验在一台超精密的车床上进行，该车床在平移和转动运动中均具有良好的刚度和稳定性。

50. conductivity ［ˌkɑːndʌkˈtɪvəti］ n. 传导性

例句1 Copper C110 was selected as the foil material due to its high **electrical conductivity**.

C110铜因其较高的导电性而被选作箔材料。

例句2 Machining of nickel alloys is difficult due to a combination of high temperature strength and low **thermal conductivity**.

高温强度和低导热性的两者结合，给镍合金的加工带来困难。

51. configuration ［kənˌfɪɡjʊˈreɪʃn］ n. 配置

例句1 This **assembly configuration** can be obtained by overlapping the links a34 with a3040 and the links a4010 and a400100.

这种装配配置可以通过将连杆a34与a3040以及连杆a4010和a400100重叠获得。

例句 2　The selection and assembly of specific components to accomplish a well-defined objective is a familiar task in engineering, commonly referred to as **configuration design**.

为完成一个明确的目标而选择和装配特定的部件是工程中常见的任务,通常被称为配置设计。

52. consistent　[kənˈsɪstənt]　adj. 一致的

例句 1　These values remained relatively **consistent for** powder bed temperatures from 100 to 670 ℃.

当粉床温度处于 100~670℃时,这些值保持相对一致。

例句 2　It was experimentally determined that the discharging delay is approximately 10 ms, which is **consistent with** the high speed camera imaging result.

实验结果表明,放电延迟大约为 10 ms,这与高速相机成像结果一致。

53. consists　[kənˈsɪsts]　v. 由……组成(单数,第三人称,一般现在时)

例句 1　The workpiece **consists of** two bearing rings, denoted as the top bearing ring and bottom bearing ring.

工件由两个轴承环组成,分别表示为上轴承环和下轴承环。

例句 2　This table **consists of** two orthogonal axes that are driven by high performance brushed DC servo motors through low friction ball screws.

该工作台由两个正交的轴组成,它们由高性能的有刷直流伺服电动机通过低摩擦滚珠丝杠驱动。

54. constraints　[kənˈstreɪnts]　n. 约束(复数)

例句 1　At the beginning, the strategy is more conservative by individually modelling all inequality **constraints**.

起初,该策略更为保守,它单独对所有不等式约束进行建模。

例句 2　The geometry of the design often dictates which process will be most capable of creating the part due to **manufacturing constraints**.

由于制造方面的限制,设计的几何形状往往决定了哪种工艺最能制造出该零件。

55. constructed　[kənˈstrʌktɪd]　v. 构建(过去分词)

例句 1　The workpiece and tool chains are **constructed from** the base coordinate frame to the workpiece and tool, respectively.

工件链和刀具链是分别从基础坐标系到工件和刀具构建的。

例句2　The result indicates that a family of overconstrained mechanisms can be **constructed by** combining legitimate modules.

结果表明，通过组合合理的模块，可以构建一组过约束机制。

56. construction　［kənˈstrʌkʃn］　n. 建造；构建

例句1　Perception and cognition play important roles in this **construction of** preference.

感知和认知在这种偏好的构建中发挥着重要作用。

例句2　Additive manufacturing is currently also used for the **construction of** end use components.

增材制造在目前也被用于最终用途组件的建造。

57. contrast　［ˈkɑːntræst］　n. 对比

例句1　This is **in contrast to** prior findings showing complexity did not impact learning during product dissection.

这与之前的研究发现形成了鲜明对比，之前的结果显示复杂性并不会影响产品剖析过程中的学习。

例句2　**By contrast**, in dry cutting, since the temperature is very high and a great deal of energy is introduced into the microstructure, partial dynamic recrystallisation occurs.

相比之下，在干切削中，由于温度非常高，大量的能量被引入到微观结构中，因此会发生部分动态再结晶。

58. controllers　［kənˈtrəʊləz］　n. 控制器（复数）

例句1　Three conventional **closed-loop controllers** are also implemented and assessed their torque tracking performances.

三种传统的闭环控制器也被应用，并评估了它们的扭矩跟踪性能。

例句2　The presence of the small error in the constant speed linear portion of the path is below the tolerance limit, but it indicates that the **axis controllers** are not perfectly matched.

在路径的恒速线性部分中存在的小误差低于公差极限，但表明轴控制器并未完全匹配。

59. create ［kriˈeɪt］ v. 创造

例句 1　For example, a prototype can first **create** a common language between two or more people.

例如，一个原型可以首先在两个或更多的人之间创建一种通用语言。

例句 2　Therefore, reusing information can **create** unprecedented opportunities in advancing the theory, method, and practice of product design.

因此，信息再利用可以为推进产品设计的理论、方法和实践创造前所未有的机遇。

60. created ［kriˈeɪtɪd］ v. 创造（过去分词）

例句 1　Further, the binary tree structure was **created by** the research team and the organization of the questions was designed to minimise the number of strategies.

此外，二叉树结构是由研究团队创建的，问题的组织设计是为了使策略的数量最小化。

例句 2　The test revealed that the functionality of designs **created by** the conventional group did not differ from the functionality of designs produced by the additive group.

测试表明，由常规组创建的设计的功能性与由添加组生成的设计的功能性没有区别。

61. creating ［kriˈeɪtɪŋ］ v. 创造（动名词）

例句 1　One of the purposes of **creating** products for developing countries is to improve the consumers' quality of life.

为发展中国家创造产品的目的之一是改善消费者的生活质量。

例句 2　The students showcase their learning by **creating** a novel toy then manufacturing it by the end of the semester.

学生们创造一种新奇的玩具，并在期末制造出来，以展示他们的学习成果。

62. curvature ［ˈkɜːvətʃə］ n. 曲率

例句 1　This prevents the stress induced **curvature** of the reflectors.

这可以防止应力引起的反射体曲率。

例句 2　The joints must bend through large angles with a small radius of **curvature** without failure.

接头必须在不发生故障的前提下以较小的曲率半径大角度弯曲。

63. cycle　　[ˈsaɪkəl]　n. 周期

例句 1　It is observed that the **cycle time** after implementation of smoothing algorithm is 2.849s.

可以看出，实施平滑算法后的周期时间为 2.849 s。

例句 2　The intention of the rather long **duty cycle** was to check for response variations due to environment fluctuations or creep.

较长的占空比旨在检查由于环境波动或蠕变造成的响应变化。

64. decomposition　　[ˌdiːkɒmpəˈzɪʃn]　n. 分解

例句 1　Li proposed **system decomposition method** using matrix-based two-phase approach.

Li 提出了基于矩阵两阶段法的系统分解方法。

例句 2　Table 4 highlights the effects of the three in-situ martensite **decomposition techniques** discussed.

表 4 突出了所讨论的三种原位马氏体分解技术的效果。

65. define　　[dɪˈfaɪn]　v. 定义

例句 1　The tangent vectors (*t*, *n* and *e*) **define** the orientation of each differential edge segment.

切线向量(*t*、*n* 和 *e*)定义了每个微分边缘部分的方向。

例句 2　The greater the number of points used to **define** a tool path, the greater the processing speed required from the controller.

用于定义刀具路径的点的数量越多，对控制器的处理速度要求就越高。

66. deformation　　[ˌdiːfɔːˈmeɪʃn]　n. 变形

例句 1　Investigation of **material deformation** at the chisel edge is required for complete simulation of drilling processes.

为了完全模拟钻孔过程，需要对横刃材料的变形进行研究。

例句 2　Therefore, it appears that the high temperatures are leading to either localized melting or **plastic deformation** of the abrasive.

因此，高温似乎能导致磨料的局部熔化或塑性变形。

67. demonstrate　　[ˈdɛmənˌstreɪt]　v. 表明，演示

例句1　These results demonstrate that the worksheet can help reduce the design cycle for novice and intermediate users.

这些结果表明,工作表可以帮助新用户和中级用户缩短设计周期。

例句2　In the rest of the paper, we demonstrate how the conditions can be applied to robust design optimization technique.

在本文的余下部分,我们将展示如何把这些条件应用于稳健设计优化技术。

68. demonstrated ['demənstreɪtɪd] v. 表明(过去分词)

例句1　It will be demonstrated that the generated path can be processed by the machine with a higher average feed rate.

实验结果表明,所生成的路径能够通过较高的平均进给速度的机床进行加工。

例句2　The shear plane heat source was demonstrated to be more significant than the flank face frictional heat source on peak machined surface temperature.

剪切面热源比后刀面摩擦热源对加工表面峰值温度显示出更大的影响。

69. dense [dɛns] adj. 稠密的

例句1　The powder material is completely melted and solidified with an aim to achieve fully dense parts.

粉状物料完全熔化和凝固,以获得致密零件。

例句2　Furthermore, the energy density variable has been commonly used as a means to define a process window for fully dense components.

此外,能量密度变量通常是用来定义致密零件工艺窗口的手段。

70. density ['dɛnsɪtɪ] n. 密度

例句1　They found that the packing density (or porosity) significantly affects the temperature profile.

他们发现装填密度(或孔隙率)对温度曲线影响显著。

例句2　A common method to attempt to quantify the presence of defects within components is using the energy density variable.

试图量化部件内缺陷的一种常用方法是使用能量密度变量。

71. deposition [,depə'zɪʃn] n. 沉积

例句1　Thus, orientation control is also necessary to achieve consistent deposition of good quality glass.

因此，为了实现优质玻璃的一致沉积，方向控制也是必要的。

例句2 In addition, designers should also pay attention to grain size of thermoresistors in choosing **deposition temperature** and thickness of thermoresistors.

此外，设计者在选择热敏电阻的沉积温度和厚度时还应注意热敏电阻的晶粒大小。

72. derived ［dɪˈraɪvd］ v. 导出，得出（过去分词）

例句1 All other clothoid equations are **derived from** equation (1).

所有其他回旋曲线方程都是由方程(1)导出的。

例句2 The total volume of gas generated can therefore be **derived by** assuming all bubbles around the tool electrode have similar properties.

因此，通过假设工具电极周围的所有气泡都具有类似的特性，可以推导出产生的气体总体积。

73. designed ［dɪˈzaɪnd］ v. 设计（过去分词）

例句1 This device was **designed to** be easily manually deployed with limited localization guidance.

该设备被设计成在有限本土化指导下易于手动部署。

例句2 The system is **designed to** carry out steering actions that would otherwise be carried out manually by the surgeon.

该系统被设计用来执行原本由外科医生手动执行的操作步骤。

74. designs ［dɪˈzaɪnz］ n. 设计（复数）

例句1 Then, they were asked to modify their **designs** and make them easier to machine.

然后，他们被要求修改他们的设计，使其更易于加工。

例句2 **Designs** with a low buy-to-fly ratio are expected to be classified as cast parts.

低买飞比的设计将被归类为铸造零件。

75. detection ［dɪˈtekʃn］ n. 检测

例句1 Another application of the thermoresistive effect is the **detection of** acceleration.

热阻效应的另一个应用是对加速度的检测。

例句2 Sensitive thermoresistive elements are commonly employed for the **detection of** temperature changes, which come from the velocity or direction change of fluids.

敏感的热阻元件通常被用于检测温度变化，这种变化源于流体的速度或方向的改变。

76. deviation ［ˌdiːviˈeɪʃn］ n. 偏差

例句1 As the flow rate increases, the **standard deviation of** the average pressure measurements also increases.

随着流速的增加，平均压力测量的标准偏差也在增加。

例句2 The **straightness deviation of** deep holes is the most important performance indicator in gun drilling process.

深孔的直线度偏差是深孔加工工艺中最重要的性能指标。

77. device ［dɪˈvaɪs］ n. 装置，设备

例句1 A single grit pullout **device** is developed and utilized for this analysis.

研发了单磨粒脱离装置并用于此分析。

例句2 The intended application of the **device** is analyte mass sensing and to identify on which resonator the analyte has docked.

该装置的预期应用是分析物质量传感并识别分析物停靠在哪个谐振器上。

78. devices ［dɪˈvaɪsɪz］ n. 装置，设备（复数）

例句1 These **devices** are the smallest ever reported microhydraulic systems that include the actuation source.

这些设备是有史以来报道过的包含驱动源的最小的微液压系统。

例句2 **Devices** are usually implemented on the front side of the folded structure to provide an easy access for inspection and probing.

设备通常被安装在折叠结构的正面，以便为检查和探测提供便捷通道。

79. differential ［ˌdɪfəˈrenʃl］ adj. 微分的

例句1 The initial value of each technology performance is also required to solve **differential equations**.

也需要每项技术性能的初始值来求解微分方程。

例句 2　It is an algebraic model, except the temperature-electrical input relationship, which is put forward as a **differential equation**.

除温度-电输入关系以微分方程的形式提出外,它是一个代数模型。

80. digital　[ˈdɪdʒɪtl]　adj. 数字的,数码的

例句 1　Thus, the measurements for the piston movement had to be made using only the **digital camera** and a tracking software, and could not be automated.

因此,活塞运动的测量只能通过数码相机和跟踪软件进行,无法自动化。

例句 2　This procedure significantly increases the **digital file** size of the unit cell based on the overall triangle count (approximately 50,000/unit cell).

这个程序显著地增加了基于整个三角形计数单元格的数字文件大小(大约每个单元 50 000)。

81. dimensional　[dɪmenʃənl]　adj. 尺寸的

例句 1　**Dimensional synthesis** aims at finding the optimal linkage parameters.

尺寸合成的目的在于寻找最佳的连接参数。

例句 2　Three thick rods are included on the top of the part for evaluating **dimensional accuracy** using a 3D scanner.

部件顶部有三根粗杆,用以借助三维扫描仪评估尺寸精度。

82. dimensions　[daɪˈmenʃ(ə)nz]　n. 尺寸;维度(复数)

例句 1　The **optimal dimensions** are given in Table 1.

表 1 中给出了最佳尺寸。

例句 2　With six **dimensions of** strain, three dimensions of field, and many modes, an analytical solution is not possible.

对于应变的 6 个维度、场的 3 个维度和许多模态,解析解是无法实现的。

83. displacement　[dɪsˈpleɪsmənt]　n. 位移

例句 1　The tests were repeated for different load resistance and initial tip **displacement** values.

对不同的负载阻力和初始刀尖端位移值进行了重复测试。

例句 2　The laser **displacement** sensor is placed underneath the actuator in an inverted orientation.

激光位移传感器被倒置于执行器的下方。

84. distributed ［dɪˈstrɪbjuːtɪd］ v. 分布，分散（过去分词）

例句1 Since the loads are **evenly distributed** across all of the pins, a lower clamping force can generally be permitted.

由于载荷均匀地分布在所有的销上，因此通常情况下夹紧力可以小一些。

例句2 For the uniform pressure theory, the pressure is assumed to be **uniformly distributed** over the frictional surface.

根据均匀压力理论，压力是均匀分布在摩擦表面上的。

85. distribution ［ˌdɪstrɪˈbjuːʃn］ n. 分布

例句1 **Distribution of** the electric field does not experience any significant variation either.

电场的分布也没有明显的变化。

例句2 The **distribution of** contact stresses along the rake and flank tool faces is shown in Fig. 20.

图20显示了沿前刀面和后刀面的接触应力分布。

86. domain ［dəʊˈmeɪn］ n. 域

例句1 Roukema and Altintas presented a **time domain simulation** of drilling by considering the exact kinematics of the process.

基于过程的精确运动学，Roukema 和 Altintas 提出了一种钻井的时域模拟。

例句2 In this paper, this **frequency domain analysis** is used to improve the choice of the target profile in such a way that it can be etched.

在本文中，这种频域分析被用来改进目标轮廓的选择，使其能够被蚀刻。

87. dynamic ［daɪˈnæmɪk］ adj. 动态的

例句1 Fig. 6 represents grinding as a **dynamic system**.

图6将研磨显示为一个动态系统。

例句2 The theoretical sensitivity and **dynamic range of** the sensor are also analyzed in this section.

本节还分析了该传感器的理论灵敏度和动态范围。

88. electron ［ɪˈlektrɒn］ n. 电子

例句1　The heat penetration of **electron beam** is higher than laser melting processes.

电子束的热穿透性能要优于激光熔化过程。

例句2　To investigate this, the wear spot in a worn single-point diamond was analyzed under an **electron microscope** (Fig. 13).

为了研究这个问题，我们在电子显微镜下分析了磨损的单点金刚石中的磨损斑（见图13）。

89. element ［ˈelɪmənt］ n. 元素

例句1　The slot length and width are numerically optimized using a finite **element** method (FEM) solver.

通过有限元法（FEM）求解器对槽的长度和宽度进行了数值优化。

例句2　Cao et al. presented a finite **element** model of the spindle system and simulations were compared against experimental results.

Cao等人提出了一个主轴系统的有限元模型，并将模拟结果与实验结果进行了比较。

90. embedded ［ɪmˈbedɪd］ v. 嵌入（过去分词）

例句1　Based on our results, future work should also focus on computer-based recommender systems **embedded in** CAD (computer-aided design).

基于我们的研究结果，未来还应该关注嵌入CAD（计算机辅助设计）中的基于计算机的推荐系统。

例句2　Assembly level design freedom is mainly comprised of articulated mechanisms, functionally graded material, **embedded components**, multifunctional design, and compliant mechanisms.

装配水平的设计自由度主要由铰接机构、功能分级材料、嵌入式组件、多功能设计和柔性机构组成。

91. empirical ［emˈpɪrɪkəl］ adj. 实证的，以经验为依据的

例句1　Researchers have used **empirical studies of** decision-making to understand how human designers decompose problems.

研究人员利用决策的实证研究来探析人类设计师是如何分解问题的。

例句 2　The majority of these use regression analyses to fit a set of experimental observations to an **empirical model**.

其中大多数使用回归分析将一组实验观察结果拟合到一个经验模型中。

92. enable　[ɪnˈeɪbəl]　v. 使能够

例句 1　These positive outcomes would **enable** the optimization of process parameters.

这些正向的结果能够优化工艺参数。

例句 2　The resulting data will **enable us to** construct an ideal sintering schedule.

所得数据使我们能够构建一个理想的烧结时间表。

93. ensure　[ɛnˈʃʊə]　v. 确保

例句 1　It is important to **ensure that** the shifts in the individual eigenfrequencies are larger than the frequency noise during measurement.

重要的是在测量过程中要确保各个特征频率的偏移大于测量时的噪声频率。

例句 2　In order to **ensure that** the threads would provide a complete seal during use, an outer weld bead would be added to the outer edge of the interfaces.

为了确保在使用过程中螺纹能提供完整的密封，将在接口的外边缘增加一道外焊缝。

94. environment　[ɪnˈvaɪrənmənt]　n. 环境

例句 1　Moreover, the fabrication of thermal flow sensors have involved various solvents and chemicals which are unfriendly for the **environment**.

此外，热流传感器的制造需要用到溶剂和化学制品，这对环境是有害的。

例句 2　If the critical voltage in a specific **machining environment** is too high, machining quality can be highly affected due to high energy of the discharge.

如果特定加工环境中的临界电压过高，放电产生的高能量会使加工质量受到很大影响。

95. environments　[ɪnˈvaɪrənmənts]　n. 环境（复数）

例句 1　This kept the audio communication in both **environments** comparatively similar and eliminated confounding factors in the analysis.

这使两种环境中的声频通信保持相对相似，并在分析中消除了干扰因素。

例句 2　**Virtual environments** also allow students to interact with products and items that they would not normally interact with in a physical classroom.

虚拟环境还允许学生与他们通常在现实教室中接触不到的产品和物品进行互动。

96. equivalent　[ɪˈkwɪvələnt]　adj. 相等的，相当的

例句 1　This is **equivalent to** machine dynamic stiffness switching from positive to negative.

这相当于机器的动态刚度由正转为负。

例句 2　This step is **equivalent to** fixing the distance between the revolute joint axes Z1 and Z300.

这一步相当于固定转动关节轴 Z1 和 Z300 之间的距离。

97. established　[ɪˈstæblɪʃt]　v. 确立，建立（过去分词）

例句 1　Regarding heat treatment, some general details can **be established**.

关于热处理，有些一般细节是可以确定的。

例句 2　In this study, a mathematical model has **been established for** the material removal mechanism with relation to edge radius effect.

此项研究建立了一个与刀尖圆角半径效应有关的材料去除机制的数学模型。

98. estimate　[ˈestɪmeɪt]　v. 估计

例句 1　This value was then used to **estimate** the resonator turnover temperature.

然后用这个值来估计谐振器的反转温度。

例句 2　The grit loads were calculated from the combination of the process kinematics and the grinding power to **estimate** the resulting stress.

结合工艺运动学和磨削功率来计算磨粒载荷，用以估计产生的应力。

99. estimated　[ˈestɪmeɪtɪd]　v. 估计（过去分词）

例句 1　There is small variation in the **estimated values of** tangential coefficients.

切向系数的估计值变化不大。

例句 2　As mentioned in the modelling section, the **estimated parameters** have error up to 10%.

正如在建模部分所提到的，估计的参数有 10% 的误差。

100. estimation [ˌɛstɪˈmeɪʃn] n. 估计

例句 1 Internal sensors and kinematic information enable the **estimation of** slave position and orientation.

内部传感器和运动学信息能够估算出从属的位置和方向。

例句 2 Example 3 investigates the influence of dimensionality on the performance of **reliability estimation**.

例 3 研究了维度对可靠性估计性能的影响。

101. etched [etʃt] v. 蚀刻(过去分词)

例句 1 This approach gives information about what kind of surfaces can be **etched**.

这种方法提供了关于什么类型的表面可以被蚀刻的信息。

例句 2 The experimental tests discussed in Section 4 are all for straight paths **etched by** a jet with a variable feed speed.

在第 4 节中讨论的实验测试都是针对可变进给速度射流侵蚀的直线路径的。

102. etching [ˈɛtʃɪŋ] v. 蚀刻(现在分词)

例句 1 The centreline of these trenches can be treated as the outcome of a one-dimensional **etching process**.

这些沟槽的中心线可以被视为一维刻蚀工艺的结果。

例句 2 The Fourier transform of the **etching rate** function calculated in the previous section is also displayed (dotted black line).

上一节中计算的蚀刻速率函数的傅里叶变换也被显示出来(黑色虚线)。

103. evaluate [ɪˈvæljʊeɪt] v. 评估，估计

例句 1 We **evaluate** our methodology using relevance defined in the following equation.

我们通过以下公式中定义的相关性来评估我们的方法。

例句 2 Ren et al. clamped a thermocouple between the tool and shim to remotely **evaluate** tool-chip temperature.

Ren 等人在刀具和垫片之间夹了一个热电偶来远程评估刀具片的温度。

104. evaluated [ɪˈvæljʊeɪtɪd] v. 评估，估计(过去分词)

例句 1 The accuracy of the gaps can be **evaluated by** measuring the gaps with calipers and comparing to the intended gap width.

通过使用卡尺测量间隙并与预期间隙宽度进行比较，可以估算出间隙的精度。

例句2　Piezo electric conversion performances of the fabricated cantilevers were **evaluated by** bending and releasing the cantilever.

通过弯曲和释放悬臂来评估所制备悬臂梁的压电转换性能。

105. evaluation　[ɪˌvæljuˈeɪʃn]　n. 评价，评估

例句1　The uncertainty in defining quality may be due to uncertain **evaluation criteria** used by the sponsor.

界定质量的不确定性可能是由主办方使用的不确定的评估标准所导致的。

例句2　For a comprehensive **evaluation of** the presented approach, we therefore use a large diverse set of test problems.

为了全面评价所提出的方法，我们使用了大量不同的测试题。

106. evolution　[ˌiːvəˈluːʃn]　n. 演变

例句1　Fig. 3 shows the **evolution of** the distribution of the grit protrusion heights.

图3显示了磨粒突起高度分布的演变。

例句2　Of note, **technology evolution prediction** is also called product performance prediction because any product can be regarded as a technology.

值得注意的是，技术演变预测也被称为产品性能预测，因为任何产品都可以被看作一种技术。

107. expansion　[ɪkˈspænʃn]　n. 扩张，膨胀

例句1　Once the coolant stream enters the V-channel, it becomes even more rapid yet due to the greater **expansion of** the transport passage.

一旦冷却剂流进入V形通道，由于运输通道的延展，冷却剂流的流速变得更快。

例句2　The close similarity in the coeffcient of thermal **expansion for** Ti and TiB reduces the potential for residual stresses in the final composite.

钛和硼化钛的热膨胀系数相似，这减少了最终复合材料中出现残余应力的可能性。

108. experimentally　[ɪkˌsperɪˈmentəli]　adv. 实验上地

例句1　Based on the results of the simulation, basic nozzle designs were implemented and **experimentally tested**.

基于模拟实验的结果，进行了基本的喷嘴设计和实验测试。

例句 2　It was **experimentally determined** that the discharging delay is approximately 10 ms, which is consistent with the high speed camera imaging results.

经实验确定，放电延迟约为 10 ms，这与高速摄像成像结果一致。

109. external　[ɪkˈstɜːrnl]　adj. 外部的

例句 1　However, existing methods are not focused on the specific challenge of adapting to an internal fault or an **external** disturbance.

然而，现有的方法并不是为了应对如何适应内部故障或外部干扰的具体挑战。

例句 2　They result from the deformation and expansion of the machine structure due to temperature variation caused by internal and **external** heat sources.

它们是由于内部和外部热源引起温度变化而导致机器结构的变形和膨胀。

110. extracted　[ɪkˈstræktɪd]　v. 提取（过去分词）

例句 1　Free surface data were **extracted from** a transverse line passing through the origin.

自由表面数据是从穿过原点的横线中提取出来的。

例句 2　The **extracted parameters** showed us that the charging rate in all measurements were unaltered.

提取的参数表明，所有测量中的充电率都没有改变。

111. fabricated　[ˈfæbrɪkeɪtɪd]　v. 制造（过去分词）

例句 1　Near fully dense parts were **fabricated** using the standard process with a few modifications.

通过使用稍加修改的标准工艺制造出了接近致密的零件。

例句 2　The cantilever was **fabricated** as a technology demonstrator, which shows the magnetic properties of the film.

制作了悬臂梁用于技术演示，展示了薄膜的磁性。

112. fabrication　[ˌfæbrɪˈkeɪʃn]　n. 制造

例句 1　Section 3 presents the novel design and **fabrication of** the integrated system.

第3节介绍了集成系统的新设计和制造。

例句2　Another advantage of the simple **fabrication process** is the possibility of creating multiple layer devices.

该简单制造工艺的另一个优点是可以创建多层器件。

113. factors　[ˈfæktəz]　n. 因素(复数)

例句1　All these **factors** are predicted to adversely affect tool life.

据预测，所有这些因素都会对刀具寿命产生不利影响。

例句2　The four **factors** are material choice, orientation of the test part within the build chamber, location of the test part within the build chamber, and machine identity.

这4个因素是材料的选择、测试部件在成型室中的方向、位置和机器本身。

114. feasibility　[ˌfiːzəˈbɪləti]　n. 可行性

例句1　All the three rules are focused on checking the **feasibility of** component-pair link relation.

所有这三条规则都着重于检查组件对连接关系的可行性。

例句2　A number of experiments with different workpiece materials are run to investigate the **feasibility of** tool wear monitoring using this method.

用不同的工件材料进行的一系列实验旨在研究这种方法对刀具磨损监测的可行性。

115. feasible　[ˈfiːzəbl]　adj. 可行的

例句1　The **feasible design space** is often defined by constraints.

可行设计空间通常由约束条件来确定。

例句2　Specifically, we use the Branin function as an indicator of whether a sample is inside the **feasible domain**.

具体来说，我们使用Branin函数作为指标，确定样本是否在可行域内。

116. feature　[ˈfiːtʃə(r)]　n. 特征

例句1　The most interesting **feature of** the optimal current profile shown in Fig. 3 is the absence of negative current flow.

图3所示为最佳电流剖面，其最有趣的特征就是没有负电流的流动。

例句2　One notable **feature of** linkage design is that they enable driving several

DOFs (degrees of freedom) with fewer motors.

联动设计的一个显著特点就是它们能够用较少的电动机驱动多个自由度。

117. features ['fiːtʃə(r)z] n. 特征（复数）

例句 1　Morphological details of the end-of-life wheel are shown in Fig. 5 with different **features of** the accumulated.

报废车轮的形态细节如图 5 所示，带有所累积的不同特征。

例句 2　The high resolution of the profilometer revealed these small scale, high frequency **features of** the machined kerf.

轮廓仪的高分辨率揭示了机加工切缝小尺度及高频率的特征。

118. feedback ['fiːdbæk] n. 反馈

例句 1　The strain **feedback from** this approach was reported to be mostly linear and highly repeatable, but suffered from some hysteresis at higher strain rates as this material is allowed to refill the micro-channels.

据报道，这种方法的应变反馈主要是线性的和高度可重复的，但当这种材料重新填充微通道时会在较高的应变速率下存在一些滞后现象。

例句 2　In order to evaluate the quality and repeatability of the **feedback from** the embedded flex sensors, a soft actuator sample was repeatedly actuated at different magnitudes and durations of the input pneumatic supply.

为了评估嵌入式柔性传感器反馈的质量和可重复性，需要在输入气动供应的不同幅度和持续时间下反复驱动软执行器样品。

119. finally ['faɪnəli] adv. 最后

例句 1　**Finally**, the force model coefficients are estimated as described in Section 2.2 and used in the statistical methods of Section 3.4 to investigate whether the system is out of control.

最后，按 2.2 节所描述的那样对力模型系数进行估算，并用于 3.4 节的统计方法中，以调查系统是否失控。

例句 2　Thus, using a second-order adjoint approach, the cost of computing the Hessian matrix, and **finally** the probability of failure, is proportional to the number of random parameters.

为此，使用二阶伴随方法，Hessian 矩阵的成本计算以及最终失败概率都与随机参数的数量成正比。

120. finite ['faɪnaɪt] adj. 有限的，有限制的

例句1　**Finite element methods** have also been used to model overlapped footprints.

有限元方法也用于重叠足迹的建模。

例句2　Although some researchers have performed **finite element analysis** to the process, the relation between process parameters and machining outcomes is not clear in many cases.

尽管一些研究人员已经对工艺进行了有限元分析，但在许多情况下，工艺参数与加工结果之间的关系并不明确。

121. flexibility [ˌfleksəˈbɪləti] n. 灵活性

例句1　Its **flexibility** is modeled in axial torsional directions.

其灵活性是在轴线扭转方向建模。

例句2　Control wheel **flexibility** was considered as attenuation of regenerative feedback due to contact length.

控制轮的灵活性被认为是由于接触长度而导致的再生反馈的衰减。

122. focus [ˈfoʊkəs] v. 集中，关注；聚焦

例句1　Here we **focus on** the environmental pillar.

在这里，我们专注于环境支柱。

例句2　Future testing will **focus on** an improved understanding of the viscous damping coefficient required to compensate for cable-based energy dissipation.

未来的测试将侧重于更好地了解补偿基于电缆的能量耗散所需的黏性阻尼系数。

123. focused [ˈfoʊkəst] v. 集中，关注；聚焦（过去分词）

例句1　Much of this research has **focused on** how virtual tools can impact efficiency, or the speed and effectiveness of task completion.

大部分的研究都集中在虚拟刀具如何影响效率或任务完成的速度及有效性上。

例句2　The previous section has **focused on** the modelling of tool and chip geometry, kinematics of chip generation and corresponding cutting forces with and without vibrations; these are the foundations of cutting dynamics model.

上一节的重点是刀具和切屑几何形状的建模、切屑生成的运动学以及有振动和无振动的相应切削力；这都是切削动力学模型的基础。

124. framework　［ˈfreɪmwɜːrk］　n. 结构、框架

例句 1　The **moulding framework** is highlighted in the flowchart of Fig. 9.

图 9 的流程图中着重标记了建模框架。

例句 2　The **modelling framework** is also utilized to infer electroplated CBN grinding wheel life expectancy for traditional high speed grinding processes.

建模框架还用于推断传统高速磨削工艺中电镀立方氮化硼砂轮的预期寿命。

125. frequencies　［ˈfriːkwənsiz］　n. 频率(复数)

例句 1　Negative up boundaries occur close to **frequencies of** geometric instability.

负向上边界发生在接近几何不稳定的频率上。

例句 2　The fundamental resonance **frequencies of** micro-devices are typically higher than their macro-scale counterparts by orders of magnitude.

微器件的基本谐振频率通常比宏观尺度的对应器件高出几个数量级。

126. functions　［ˈfʌŋkʃnz］　n. 功能、函数(复数)

例句 1　They have two time delays which are **functions of** spindle speed and orbital motion.

它们有两个时间上的延迟，分别是主轴速度和轨道运动的函数。

例句 2　Combining nose grind contour with flank face geometries for relief purposes brings about the profile and shape **functions of** the bottom clearance.

将刀尖磨削轮廓与后刀面几何形状相结合达到后凸目的时，可产生底部间隙的轮廓和形状功能。

127. furthermore　［ˌfɜːrðərˈmɔː(r)］　adv. 此外，而且

例句 1　**Furthermore**, the separation between each indent should be no less than 2.5 times the indenter size to avoid edge-softening effects.

此外，每个压痕之间的间隔应不小于压头尺寸的 2.5 倍，以避免边缘软化效应。

例句 2　**Furthermore**, the abrasive action is likely to be of a progressive nature due to the varying locations of the abrasives around the tool's periphery and their heights of protrusion from the containing bond.

此外，磨料作用可能具有渐进性质，因为磨料在砂轮外围不同的位置，并且它们在黏结剂中突出的高度也不同。

128. generate ［ˈdʒenəreɪt］ v. 产生，引起

例句 1　In order to **generate** smooth high-speed motion along the tool path with linear segments, global smoothing approach was employed by Yuen et al.

为了沿线性段刀具路径产生平滑高速运动，Yuen 等人采用了全局平滑方法。

例句 2　Fig. 11(a) shows the axis velocity and acceleration profiles of the pre-compensated reference commands used to **generate** the simulation and experimental results of Figs. 6-8.

图 11(a) 显示了用于生成图 6 至图 8 中仿真和实验结果的预补偿参考命令的轴速及加速度曲线。

129. generation ［ˌdʒenəˈreɪʃn］ n. 产生；(产品发展的) 代

例句 1　This slowly increases the edge blunting of the tool which ultimately leads to the **generation of** high cutting forces leading to progressive tool failure.

这会慢慢增加刀具的刃口钝化，最终导致高切削力的产生，并造成渐进式刀具失效。

例句 2　However, the current **generation of** industrial fixture sensing is somewhat rudimentary in its application with examples including "go-no-go" sensors or variant detection to ensure the correct component has been loaded into the fixture.

然而，当前一代工业夹具传感还处于其应用的初级阶段，通过"go-no-go"传感器或变体检测来确保正确的组件能加载到夹具中。

130. geometric ［ˌdʒiːəˈmetrɪk］ adj. 几何的

例句 1　The **geometric set-up** chosen and this natural frequency are a poor combination.

所选的几何设置和这个固有频率组合起来并不理想。

例句 2　Since the **geometric errors** are not effected by the feed rate, the residual errors are likely to be caused by dynamic errors, as those are feed rate influenced.

由于几何误差不受进给率的影响，因此残余误差很可能是由动态误差引起的，

因为后者受进给率的影响。

131. geometry ［dʒiˈɑːmətri］ n. 几何；几何形状

例句 1　Therefore, the direction of the resultant force on the tool stays close to what is expected based on the **geometry of** the cut.

因此，刀具上产生的合力方向同基于切削几何形状的预期方向接近。

例句 2　The key findings can be summarised as follows: the 2D profiles of the scratch marks replicate the envelope **geometry of** the grit.

可以将主要发现总结如下：划痕的 2D 轮廓复制了磨粒的包络几何形状。

132. global ［ˈɡloʊbl］ adj. 全局的

例句 1　At the end, the **global** surrogate model is built based on Eq. (4).

最后，全局代理模型基于等式(4)构建。

例句 2　Both local and **global** corner smoothing algorithms successfully smoothen the discrete path geometry and deliver a continuous motion.

局部和全局拐角平滑算法都成功地平顺了离散路径几何形状，并提供了连续的运动。

133. height ［haɪt］ n. 高；高度

例句 1　For a **build height** of 24.43 mm, the melt pool length observed through experiments was approximately 1.68 mm, 1.55 mm, 1.45 mm and 1.25 mm respectively.

对于 24.43 mm 的构建高度，通过实验观察到的熔池长度分别约为 1.68 mm、1.55 mm、1.45 mm 和 1.25 mm。

例句 2　The final **height of** the coupons is less than 15 mm, as wire electrical discharge machining is used to separate the built coupons from the platform, and some of the coupon remains attached to the platform.

试片的最终高度小于 15 mm，因为使用电火花线切割将制作好的试片与平台分离，部分试片仍然附着在平台上。

134. hence ［hens］ adv. 因此；之后

例句 1　However, these methods still require modelling of each geometry individually, **hence** the mechanics of the cutting process must also be adapted to each tool geometry and process individually.

然而，这些方法仍然需要单独对每个几何形状进行建模，因此切削过程的技术细节也必须分别适应每个刀具几何形状和过程。

例句 2 Another data-driven modelling technique investigated here is the use of a feed-forward artificial neural network (ANN) that is known to cope well with handling sources of uncertainty, and is **hence** a good candidate for modelling the complex behaviour of continuum soft robots in general.

这里研究的另一种数据驱动建模技术是使用前馈人工神经网络(ANN)。众所周知，该神经网络可以很好地处理不确定性源，因此通常是连续型软机器人复杂行为建模的优秀备选方案。

135. identical ［aɪˈdentɪkl］ adj. 完全相同的

例句 1 Coupling of **identical** resonators results in appearance of several, often closely-spaced, mode frequencies for the coupled system.

相同谐振器的耦合会导致耦合系统出现多个通常间隔很近的模式频率。

例句 2 Some observations into the characteristics of the CRA can help in finding the system roots particularly when the CRA is symmetric and made of **identical** resonators.

对CRA(目录恢复区)特性的观察有助于找到系统根目录，特别是当CRA是对称的并且是由相同的谐振器组成时。

136. identification ［aɪˌdentɪfɪˈkeɪʃn］ n. 辨认，识别

例句 1 **Identification of** the type and extent of the subsurface modification is an important aspect of any machining process.

识别次表面更改类型和程度是加工过程的一个重要环节。

例句 2 The **identification of** productive tool and workpiece spindle speeds is demonstrated using chip load limit of the tools and torque-power constraints of the turn-mill machines.

对生产刀具和工件主轴速度的识别是通过使用刀具的切屑负载限制和车铣复合机床的扭矩功率约束来体现的。

137. identified ［aɪˈdentɪfaɪd］ v. 认出，识别（过去分词）

例句 1 The process damping coefficients are **identified from** chatter-free orthogonal turning tests.

工艺阻尼系数是通过无颤振正交车削测试确定的。

例句 2　Hence, there are additional forces applied at the bottom section of the tool and these forces can be **identified as** constants.

因此，在刀具的底部施加了额外的力，这些力可以被看作常数。

138. identify　[aɪˈdentɪfaɪ]　v. 认出，识别

例句 1　Tests with different feed rates were carried out to **identify** the remaining error source.

针对不同进给率的测试可以识别其余的误差源。

例句 2　However, there are few simple and fast methods to **identify** whether the machine is in a "usable" condition.

但是，很少有简单快速的方法来识别机床是否处于"可用"状态。

139. illustrated　[ˈɪləstreɪtɪd]　v. 说明；表明；显示（过去分词）

例句 1　Centreless grinding geometry is **illustrated in** Fig. 1.

无心磨削几何形状如图 1 所示。

例句 2　The trimming cuts used in the experiments ensured that the surface was cut at a steady state, **as illustrated in** Fig. 2.

如图 2 所示，实验中使用的修整切口确保表面在稳定状态下进行切割。

140. images　[ˈɪmɪdʒɪz]　n. 图像；影像；肖像（复数）

例句 1　A digital optical microscope was utilized to obtain **images of** the electro-polished surfaces from which the melt pool width and depth were measured.

利用数字光学显微镜获得电抛光表面的图像，从中测量熔池宽度和深度。

例句 2　The stereoscopic **images of** different samples along with isometric images taken from white light interferometer have been portrayed in Fig. 12.

图 12 描绘了不同样品的立体图像以及从白光干涉仪拍摄的等距图像。

141. impact　[ˈɪmpækt]　n. 影响

例句 1　It is possible to assess the **impact of** the nozzle as a whole or any individual feature by quantifying the area under the contour of the nozzle.

通过量化喷嘴轮廓下的面积，使评估整个喷嘴或任何单独特征的影响成为可能。

例句 2　These complex nozzle tip geometries require a new metric other than the simple nozzle stand-off distance to understand the **impact of** the nozzle design on the resultant profile created.

这些复杂的喷嘴尖端几何形状需要一个新的指标，而不是简单的喷嘴间隔距离，这样才能了解喷嘴设计对创建的最终轮廓的影响。

142. implementation [ˌɪmplɪmenˈteɪʃn] n. 实施，执行

例句1 The second requirement for **implementation of** the system is a force sensor.
实现该系统的第二个要求是有一个力传感器。

例句2 For **implementation of** the tool monitoring system described above, there are several technical challenges that need to be investigated and addressed.
为了实现上述刀具监控系统，需要研究和解决几个技术难题。

143. implemented [ˈɪmplɪmentɪd] v. 实施（过去分词）

例句1 The trajectory is **implemented in** a milling process and the profile is measured.
在铣削过程中实现轨迹并测量轮廓。

例句2 In order to remedy this, a lens glare filter is **implemented in** the region close to the melt pool.
为了修补这个问题，在靠近熔池的区域安装了镜头眩光滤光片。

144. index [ˈɪndeks] n. 指数

例句1 The same log function is used in the UN human development **index** to scale the impact of increasing income.
联合国人类发展指数使用相同的日志功能来衡量收入增加的影响。

例句2 By applying the two approaches to two multi-objective problems, the efficiency of using the Pareto shape **index** for weighting objectives to identify solutions is demonstrated.
通过将这两种方法应用于两个多目标问题，有效证明了帕累托形状指数加权目标可用于解决方案的确定。

145. indicate [ˈɪndɪkeɪt] v. 表明，指示

例句1 Positive up boundaries **indicate** instability for forces larger than the threshold value.

正向上边界表明了大于阈值力的不稳定性。

例句 2　These images clearly **indicate that** the presence of gas film is required for sparks to take place.

这些图像清楚地表明产生火花时需要有气膜的存在。

146. indicates　　[ˈɪndɪkeɪts]　　v. 表明，指示（第三人称单数一般现在时）

例句 1　The model of electrolysis **indicates that** the reaction rate of generating bubbles and films is proportional to current density.

电解模型表明，产生气泡和薄膜的反应速率与电流密度成正比。

例句 2　This value of close to 40% **indicates that** the melt pool is considerably asymmetrical, indicating a strong effect of the already processed scanned hatch（or track）.

接近40%的值表明熔池相当不对称，表明对已经处理过的扫描舱口（或轨迹）的强烈影响。

147. indicating　　[ˈɪndɪkeɪtɪŋ]　　v. 表明，指示（现在分词）

例句 1　The operation labels for the emission matrix also contain a percentage **indicating** the raw frequency with which each operation occurred.

发射矩阵的操作标签还包含一个百分比，指示每个操作发生的原始频率。

例句 2　An FFT（fast Fourier transform）of the wheel while idling（c）showed much smaller values and different and somewhat higher frequencies, **indicating that** the peaks during dressing are likely coming from the grit/diamond contact.

车轮在怠速时的快速傅里叶变换算法显示出更小的值和不同且稍高的频率，表明车轮修正期间的峰值可能来自砂砾/金刚石接触。

148. induced　　[ɪnˈdjʊst]　　v. 诱发（过去分词）

例句 1　The particle size was **reduced**, and the distribution was more uniform due to increased liquid/solid wettability and dragging of the solid/liquid phase boundaries induced by the adsorption effect.

由于吸附效应引起的液/固润湿性增加和固/液相边界的拖曳，颗粒尺寸减小且分布更加均匀。

例句 2　The measurement of thermally **induced** volumetric errors has been done by several methods, including but not limited to laser interferometers, double ball bars, proximity sensors, contact displacement sensors and touch trigger probes.

热致体积误差的测量已经通过几种方法完成，包括但不限于激光干涉仪、双球杆、接近传感器、接触位移传感器和触摸触发探头。

149. instance ['ɪnstəns] n. 例子，实例

例句 1　**For instance**, when cutting conditions are held constant except for the radial immersion, there is small variation in the estimated values of tangential coefficients.

例如，当切削条件保持不变时，除了径向浸没外，切向系数的估计值变化很小。

例句 2　**For instance**, they treat geometric errors of groove profile and balls combined as a prespecified axial error term, which does not adequately capture the multidirectional interactions between groove profile errors, contact forces and elastic deformations.

例如，他们将沟槽轮廓和滚珠的几何误差视为预先指定的轴向误差项，这并不能充分捕获沟槽轮廓误差、接触力和弹性变形之间的多向相互作用。

150. integration [ˌɪntɪ'ɡreɪʃn] n. 结合，整合

例句 1　**Integration of** each axis acceleration component results in axis velocity.

每个轴加速度分量的积分会产生轴速度。

例句 2　However it highlights the potential of new research capable of improving manufacturing processes through the **integration of** conventional and additive manufacturing processes.

然而，它突出了能够通过集成传统和增材制造工艺来改善制造工艺的新研究的潜力。

151. interaction [ˌɪntər'ækʃn] n. 互动，交流；相互影响，相互作用

例句 1　This section models the **interaction between** the generalized metal cutting process and the vibrating structure.

本节模拟了广义金属切削工艺与振动结构之间的相互作用。

例句 2　It may be necessary to include a variety of sustainable attributes in a product to investigate how the **interaction between** preference elicitation methods and types of sustainable attributes affect the evaluation of a product's sustainability.

可能有必要在产品中包含各种可持续属性，以调查偏好诱导方法及可持续属性类型之间的相互作用如何影响产品的可持续性评估。

152. internal [ɪnˈtɜːrnl] adj. 内部的

例句1　The **internal pressure** response can be easily measured using common pressure sensors connected to the pneumatic supply tubes.

使用连接到气动供应管的普通压力传感器可以轻松测量内部压力响应。

例句2　In gundrilling, coolant that is supplied under high pressure travels along the **internal** conduits of drill shafts and drill tips before ejected through coolant hole(s) on the face of a gun drill to reach the drill point.

在深孔加工过程中，在高压下供应的冷却液沿着钻轴和钻头的内部导管行进，然后通过枪钻表面的冷却液孔喷射到钻点处。

153. inverse [ˌɪnˈvɜːrs] adj. 相反的

例句1　The **inverse kinematics** of the turn-milling process is modeled with the screw theory which allows a global description for rigid body kinematics.

车铣复合加工过程的逆运动学采用螺旋理论建模，该理论允许对刚体运动学进行全局描述。

例句2　The **inverse problem** is usually solved by simply controlling dwell time in proportion to the required depth of milling, without considering whether the target surface can actually be etched.

相反问题通常通过简单地控制停留时间与所需的铣削深度比例来解决，而不考虑目标表面是否可以实际蚀刻。

154. investigate [ɪnˈvestɪɡeɪt] v. 调查，研究

例句1　Finally, in order to **investigate** the limitations of the assumption of linearity, the same trajectory is repeated and the trenches allowed to overlap.

最后，为了研究线性假设的局限性，要重复相同的轨迹并允许沟槽重叠。

例句2　The main goal is to **investigate whether** these coefficients, estimated and tracked in real-time, can be used as an indirect method of estimating the tool's condition.

主要目标是研究这些实时估计和跟踪的系数是否可以用作判断刀具状态的间接方法。

155. investigated [ɪnˈvestɪɡeɪtɪd] v. 调查，研究（过去分词）

例句1　Indeed, this premature transition was shown to occur along the spiral tool paths **investigated in** section 5.

事实上，这种过早的转变被证明发生在第5节研究的螺旋刀具路径上。

例句2 Virtual reality (VR) has been **investigated** as a solution to various engineering design problems as early as 1993, and even to improve engineering communication and collaboration as early as 2000.

虚拟现实早在1993年就被尝试作为各种工程设计问题的解决方案，在2000年就已被用于改善工程沟通和协作。

156. issue ［ˈɪʃuː］ n. 问题

例句1 The zone 3 portion was not cut due to an **issue with** the controller.

由于控制器问题，区域3部分未被加工。

例句2 Moreover, this **issue** must be addressed with caution, as highly complicated mechanical configurations for the pin setting procedure reduces the attractiveness of pin-array fixtures on the workshop.

此外，必须谨慎解决这一问题，因为引脚设置程序的高度复杂的机械配置会降低引脚阵列夹具在车间的吸引力。

157. issues ［ˈɪʃuːz］ n. 问题（复数）

例句1 However, thermal **issues** and transient behavior of the grinding wheel wear directly affect the workpiece surface integrity and tolerances.

然而，砂轮磨损的热问题和瞬态行为直接影响工件表面的完整性和公差。

例句2 We propose a bracing solution to address the **issues of** providing stability, increase the potential force that can be applied to tissue, and therefore improve dexterity during catheter-based valve repair.

我们提出了一种支撑解决方案来解决稳定性提供的问题，并增加可施加到组织的潜在力，从而提高基于导管的瓣膜修复过程的灵活性。

158. lateral ［ˈlætərəl］ adj. 横向的；侧面的

例句1 This peak and the growing amplitude of the **lateral force** and cutting sound indicate lateral chatter.

这个峰值以及横向力和切割声的不断增大的振幅表明横向颤动。

例句2 Usually the long overhang of the drills in deep hole drilling, and the spiral design of the drill bits make them prone to vibration instability in **lateral** and torsional-axial directions.

通常，深孔钻头的长悬伸和钻头的螺旋设计容易使其在横向和扭转轴方向上的振动不稳定。

159. layer ['leɪə(r)] n. 层；层次

例句1 Consecutive layers are built by processing powder material with a pre-specified powder **layer** thickness.

连续层是通过加工具有预先指定粉末层厚度的粉末材料构建的。

例句2 A white **layer** having high hardness, as compared to bulk material, was reported after macro scale drilling of a nickel-base super alloy.

据报道，在对镍基超合金进行宏观钻孔后能产生相对于块状材料具有高硬度的白色层。

160. layers ['leɪə(r)z] n. 层；层次（复数）

例句1 These consecutive **layers** are processed slightly differently to ensure a robust build.

这些连续层的处理方式略有不同，以确保构建稳健。

例句2 Composition stoichiometry analysis was conducted to find out the mass percentage of the adherent **layers** of work-piece material.

进行成分化学计量分析，找出工件材料黏附层的质量百分比。

161. located ['loʊkeɪtɪd] v. 位于（过去分词）

例句1 The inserts are **located on** the tool body such that there is a bottom edge which is in contact with the workpiece.

刀片位于刀体上，以使其底刃与工件接触。

例句2 Bulk charging is identified as the charge **located on** the dielectric side connected to the metals, while surface charging is the charge generated and located at the exposed side of the dielectric.

大容量充电是位于与金属相连的电介质侧的电荷，而表面充电是产生并位于电介质暴露侧的电荷。

162. location [loʊ'keɪʃn] n. 位置

例句1 Therefore, it takes approximately 10 ms longer for the laser to reach the same x-coordinate **location** on consecutive hatches.

因此,激光到达连续舱口上相同的 x 坐标位置需要的时间大约(比原来)多出10 ms。

例句2 In agreement with Merchant's model, a narrow zone of shear originates at the **location of** the tool tip and separates the chip into lamella of nearly uniform thickness.

与 Merchant 的模型一致,狭窄的剪切区域源自刀尖的位置,并将切屑分离成厚度几乎均匀的薄片。

163. machined ［məˈʃiːnd］ v.(用机器)加工(过去分词)

例句1 The resin is used to achieve excellent edge retention at the **machined surface**.

该树脂用于在加工表面实现出色的切削刃保持力。

例句2 The mechanical and metallurgical characterisation of these surfaces is important in the aerospace industry where these surface alterations can play a key role in determining fatigue strength, stress corrosion resistance, and lifetime of the **machined component**.

这些表面的机械和冶金特征在航空航天工业中发挥重要作用,因为这些表面变化在确定疲劳强度、抗应力腐蚀性和机加工部件使用寿命方面至关重要。

164. machining ［məˈʃiːnɪŋ］ n.(用机器)加工

例句1 Comparable cutting speed tests in the **machining of** nickel-base super alloy and steel showed that the former was subjected to higher temperatures, and stresses twice those for steels, indicating higher tool wear rates.

镍基高温合金和钢加工中的可比切削速度测试显示出,前者承受的温度更高,应力是钢的两倍,表明刀具磨损率更高。

例句2 There are limitations to bringing down the process from macro to micro to ultra-precision level as several insignificant factors in **conventional machining** become prominent in ultra-precision machining.

由于传统加工中的几个微不足道的因素会在超精密加工中变得突出,因此将工艺从宏观到微观再到超精密水平降低工艺是有局限性的。

165. magnitude ［ˈmæɡnɪtuːd］ n. 规模,大小

例句1 Based on the formed surface, we may say that hatch spacing would affect the **magnitude of** surface roughness.

基于形成的表面，我们可以说舱口间距会影响表面粗糙度的大小。

例句 2　Further by constraining the **magnitude of** the resultant jerk vector, the cutting tool may follow the revised tool path exactly.

此外，通过约束所得加速度矢量的大小，切削刀具可以精确地遵循修改后的刀具路径。

166. maintain　[meɪnˈteɪn]　v. 保持，维持

例句 1　In the cutting tests, the tool overhang was kept constant to **maintain** consistent torsional stiffness.

在切削试验中，刀具悬伸保持恒定，以保持一致的扭转刚度。

例句 2　The latter occurs since a lowered bandwidth will reduce the overall path velocity to **maintain** the error limit on the particular axis concerned.

后者的发生是因为降低带宽会降低整体路径速度，以保持相关特定轴的误差限制。

167. manual　[ˈmænjuəl]　adj. 手动的；人工的

例句 1　Third, heuristic rules need **manual interpretation** and the result may vary for different designers.

第三，启发式规则需要人工解释，不同设计者的结果可能会有所不同。

例句 2　This product had 576 reviews on Amazon.com and, of those, 50 reviews were chosen for the **manual analysis**.

该产品在 Amazon 官网上有 576 条评论，其中 50 条评论被选中做人工分析。

168. methodology　[ˌmeθəˈdɒlədʒi]　n. 方法

例句 1　The **methodology** is further applied to smaller scale nozzles to investigate scalability of feature size.

该方法被进一步应用于更小规模的喷嘴，以研究特征尺寸的可扩展性。

例句 2　Therefore, future work intends to extend the **methodology** to monolithic 6-DOF parallel mechanisms and demonstrate its performance in assembly tasks.

因此，未来的研究计划将该方法扩展到单六自由度并联机构中，并展示其在装配任务中的性能。

169. micro　[ˈmaɪkroʊ]　adj. 微小的

例句 1   This type of flow is commonplace in **micro-hydraulic** systems.

这种类型的流动在微液压系统中很常见。

例句 2   Although these engineered tools might offer a set of key technological advances, up to now there is no clear understanding on what shape the **micro-cutting** edges should have.

虽然这些工程刀具可能带来一系列关键技术进步,但到目前为止,对于微切削刃应该具有什么形状尚无明确的认识。

170. microscopy　［maɪˈkrɒskəpi］　n. 显微镜

例句 1   Digital **optical microscopy** imaging and thermal camera imaging were used to corroborate these results.

数字光学显微镜成像和热像仪成像被用于证实这些结果。

例句 2   Depth profiling, **digital microscopy** and scanning electron microscopy are utilized to investigate topological evolution and mechanisms of grit failure.

深度剖析、数字显微镜和扫描电子显微镜被用于研究磨粒破坏的拓扑演化和机制。

171. minimize　［ˈmɪnɪmaɪz］　v. 使最小化

例句 1   This is to **minimize** deformation of the part during the manufacturing process.

这是为了最大限度地减少零件在制造过程中的变形。

例句 2   An average value is used to **minimize** the effect of the first calibration values, which could be high or low given the stochastic nature of cutting.

用平均值将第一个校准值的影响最小化,鉴于切削的随机性,该值可能或高或低。

172. modes　［moʊdz］　n. 模式(复数)

例句 1   The order of the **modes** in the frequency domain remains unchanged due to frequency veering.

由于频率变化,频域中模态的阶数保持不变。

例句 2   Furukawa carried out an extensive analysis of stability including two **modes** of machine vibration.

Furukawa 对机器振动的两种模式的稳定性进行了广泛的分析。

173. modified ['mɑːdɪfaɪd] v. 修改（过去分词）

例句1 The ideas were assessed based on the presence of a set of nonproducible features in initial and **modified designs**.

这些想法是根据初始和修改设计中存在的一组不可生产的特征来评估的。

例句2 Temporal position and orientation data for the **modified linear segments** are then, generated from $s(t)$ using linear interpolation technique.

然后，使用线性插值技术从 $s(t)$ 生成修改后的直线段的位置和方向数据。

174. modules ['mɔdʒulz] n. 模块（复数）

例句1 This fixture consisted of a series of **modules** containing intelligent hydraulic actuators that act as locating, supporting or clamping pins.

该夹具由一系列模块组成，其中包含智能液压执行器，如可用作定位、支撑或夹紧的销。

例句2 The performance of the locating strategy for the fixture decreases as the number of **modules** increases, due to the stack-up of tolerances for each module.

由于每个模块的公差叠加，夹具定位策略的性能会随着模块数量的增加而降低。

175. monitoring ['mɒnətərɪŋ] n. 监测

例句1 Direct tool condition **monitoring** (TCM) approaches, such as vision and optical methods, measure the geometric parameters of the cutting tool.

直接刀具状态监测（TCM）方法，比如视觉和光学方法，可测量切削刀具的几何参数。

例句2 This paper presents a new system that integrates electrohydrodynamic jet printing, an emerging micro-scale additive manufacturing technique, with inline atomic force microscopy for rapid, die-by-die inline metrology and **quality monitoring**.

本文提出了一种将电流体动力学喷射打印（一种新兴的微尺度增材制造技术）与内联原子力显微镜相结合的新系统，用于快速、逐模的内联计量和质量监测。

176. morphology [mɔːrˈfɑːlədʒi] n. 形态学

例句1 Additionally, it is found that different grain materials (Cu and Mg alloy) exhibited variations in flow stress, **chip morphology** and surface quality.

此外，还发现不同的晶粒材料（铜和镁合金）在流动应力、切屑形态和表面质量方面表现出差异。

例句2　The speed function has been discussed in more detail in literature and it has been indicated that the SF (scale factor) setting strongly affects the **surface morphology** and surface roughness.

速度函数在文献中已经进行了详细的讨论，并且已经表明比例系数(SF)设置对表面形态和表面粗糙度有很大的影响。

177. mounted　　['maʊntɪd]　　v. 安装（过去分词）

例句1　The Capball sensor is **mounted on** the machine table.

在机器工作台上安装 Capball 传感器。

例句2　Strain gages **mounted on** the tool holder body produce a signal proportional to the cutting force.

安装在刀柄上的应变片产生与切削力成比例的信号。

178. network　　['netwɜːrk]　　n. 网络

例句1　For a general **network** without community characteristics, the PCCD algorithm is fast and effective to find the minimum grouping solution.

对于没有社群特征的一般网络，PCCD 算法能快速有效地找到最小分组解。

例句2　In addition to the mechanical support of the device panels, the polymer material provides the electrical insulation for the **network of** metal traces.

除了器件面板的机械支撑外，聚合物材料还为金属走线网络提供电气绝缘。

179. nominal　　['nɒmɪnl]　　adj. 标称的

例句1　The **nominal chemical composition** is summarised in Table 1.

表1总结了标称化学成分。

例句2　A starting point is to define the **nominal** ball centre pathway of the screw, as shown by the helix in Fig. 2(a).

起点是定义螺钉的标称滚珠中心路径，如图2(a)中的螺旋线所示。

180. nonlinear　　[ˌnɒnˈlɪniər]　　adj. 非线性的

例句1　Now the **nonlinear effect** is more evident.

现在非线性效应更加明显。

例句2　The **nonlinear force characteristics** of solenoids make them hard to control and they typically have poor valve seating velocities.

电磁阀的非线性力特性使它们难以控制，并且它们通常具有较低的阀座速度。

181. normal ['nɔːrml] adj. 普通的；垂直的

例句1 The acceleration vector then lies solely **normal** to the direction of travel.

然后，加速度矢量完全垂直于行进方向。

例句2 In this case, the **normal** controller reduces path error, and the tangential controller regulates the speed that the machine tool moves along the desired path.

在这种情况下，普通控制器减少了路径误差，切向控制器调节机床沿所需路径运动的速度。

182. normalized ['nɔrməˌlaɪzd] v. 标准化；归一化（过去分词）

例句1 Because it is a **normalized** measure, the metric facilitates comparison across cue phrases that occur with different frequencies.

因为它是一种标准化的度量，所以该度量有助于比较以不同频率出现的提示短语。

例句2 Microphone measurements and their **normalized** power spectra for an unstable and chatter free stable cuts are given in Fig. 10(b) and (c), respectively.

图10(b)和(c)分别给出了用于不稳定和无颤振稳定切割的传声器测量值及其归一化功率谱。

183. novel ['nɒvl] adj. 新颖的

例句1 A **novel** design of diamond core-drill equipped with micro-teeth incorporating defined cutting edges is proposed.

一种新颖的金刚石取芯钻头设计被提出，该设计配备了包含特定切削刃的微齿。

例句2 A **novel** look-ahead windowing (LAW) technique is developed to plan tangential feed profile with uninterrupted acceleration to continuously smooth the path.

开发了一种新颖的前瞻窗口(LAW)技术来规划具有不间断加速度的切向进给曲线，以得到连续平滑的路径。

184. objective [əbˈdʒektɪv] adj. 客观的；n. 目的，目标

例句1 Pareto fronts can be inefficient due to a combinatorial increase in **objective** pairs.

目标对的组合增加会导致帕累托前沿效率低下。

例句2　The **objective** of this study is to investigate the effect of apex offset inconsistency of gun drills on hole straightness deviation in gun drilling process.

本研究旨在探讨深孔加工过程中枪钻顶点偏移不一致对孔直线度偏差的影响。

185. obtain　[əbˈteɪn]　v. 得到

例句1　To **obtain** the path error, one must first define the normal unit vector to the contour.

为了得到路径误差，必须首先定义轮廓的法向单位向量。

例句2　The oblique cutting geometry and chip area were computed based on the lip's and flute's parametric equations, and then were used to **obtain** the cutting forces and torque.

根据刃口和凹槽的参数方程计算斜刃切削几何形状和切屑面积，进而获得切削力和扭矩。

186. occur　[əˈkɜːr]　v. 发生

例句1　Selection of the initial state and range of states is influenced by the episode length and what is likely to **occur**.

初始状态和状态范围的选择受片段长度和可能发生情况的影响。

例句2　Therefore, removal at the valley floor is increased and subtle deepening of the valleys occurs polarisation effects masking areas of the profile which do not **occur** in this material.

因此，谷底的去除量增加，谷底的细微加深会发生偏振效应，从而掩盖了该材料中未发生的轮廓区域。

187. occurs　[əˈkɜːrz]　vi. 发生，出现（第三人称单数一般现在时）

例句1　As such a shift **occurs** in the rate of machining of the central feature.

因此，中心特征的加工速率会发生变化。

例句2　Like position spline, the orientation spline is also formulated in such a way that the maximum orientation error e_{max} **occurs** at the mid-point of the spline.

与位置样条一样，定向样条公式中的最大方向误差 e_{max} 也是出现在样条的中点。

188. offset ['ɔːfset] n. 偏移；抵消

例句1 The variations in dimensions led to an **offset in** stiffness.
尺寸的变化导致刚度的偏移。

例句2 Depending on diameter of the pilot hole and **offset of** the drill centre with respect to the centre of the pilot hole [X and Y in Fig. 1(b)], different portion of the cutting edge element may be engaged with the cut.
根据导向孔的直径和钻削中心相对于导向孔中心的偏移[见图1(b)中的 X 和 Y]，切削刃的不同部分可能与切口接触。

189. optical ['ɑːptɪkl] adj. 光学的

例句1 They are sufficiently fine that they appear as a featureless white layer in the **optical microscope**.
它们足够精细，在光学显微镜下看起来像无特征的白色层。

例句2 The use of an **optical polygon** with an autocollimator is able to measure the angular positioning error motion.
使用带有自准直器的光学多边形能够测量角定位误差运动。

190. optimal ['ɒptɪməl] adj. 最佳的

例句1 Hence, the **optimal** scan velocity is a trade-off between resultant density and surface quality.
因此，最佳扫描速度是所得密度和表面质量之间的权衡。

例句2 Finally, various control strategies including model reference adaptive control, fuzzy logic and genetic algorithms have been suggested for determining the **optimal** cross coupling control action.
最后，提出了各种控制策略，包括模型参考自适应控制、模糊逻辑和遗传算法，以确定最优交叉耦合控制行动。

191. optimization [ˌɒptɪməˈzeɪʃn] n. 最优化

例句1 The embrittlement observed can be related to inadequate **optimization of** the processing conditions and scanning strategy.
观察到的脆化可能与加工条件和扫描策略的优化不足有关。

例句2 In a second step the **optimization** is repeated for the measurement points B, while the previously iterated first modes in each direction are kept constant.

在第二步中,对测量点 B 重复优化,而每个方向上先前迭代的第一模式保持不变。

192. optimized ['ɒptɪmaɪzd] v. 使最优化(过去分词)

例句 1　A series of grinding tests are conducted to achieve the **optimized grinding performance** and to obtain the best cutting edge quality and grinding performance.

进行一系列磨削测试,以达到优化的磨削性能,并获得最佳的切削刃质量和磨削性能。

例句 2　It shows that the **optimized design** has a 75% larger head drop than the baseline while maintaining head values that are close to the baseline for the low and medium flow rates.

它表明,优化后设计的高差下降比基线大 75%,同时还能保持接近中低流速基线的扬程值。

193. orientation [ˌɔːriən'teɪʃn] n. 方向

例句 1　If only the position and **orientation of** the machine tool coordinate system are considered, the zero position error can be ignored.

如果只考虑机床坐标系的位置和方向,则可以忽略零位置误差。

例句 2　The position and **orientation errors** always remain within the pre-defined corner position and orientation tolerance of 0.1 mm and 0.1 degree respectively.

位置和方向误差始终保持在分别为 0.1 mm 和 0.1°的预定义角位置和方向公差范围内。

194. overall [ˌoʊvər'ɔːl; 'oʊvərɔːl] adj. 总的,整体的

例句 1　This shows that the **overall thickness** of the deformation zone (layer B) is smaller for dry cutting than wet cutting conditions, as is the extent of the finegrained region (layer A).

这表明,在干切削条件下变形区(B 层)的总厚度小于湿切削条件,细晶粒区(A 层)的厚度也是如此。

例句 2　The **overall** uncut chip **thickness** is generated by the axial rigid body motion of the drill (feed motion) and by vibrations of the drill in lateral, axial and torsional directions.

整体毛边厚度由钻头的轴向刚体运动(进给运动)和钻头在横向、轴向及扭转

方向上的振动产生。

195. overlap ［ˌoʊvərˈlæp］ n. 重叠部分

例句 1　In this case, the larger **overlap ratio** was achieved by a slower diamond traverse velocity and, in turn, a lower grit removal rate.

在这种情况下，较大的重叠率是通过较慢的金刚石横向速度实现的，进而降低磨粒去除率。

例句 2　In the medium and high **overlap** instances, the distances between variables in different subproblems are generally higher because variables are coded more often, so hierarchical clustering found the right clusters more often.

在中重叠和高重叠实例中，不同子问题中变量之间的距离通常较高，因为变量编码的频率更高，因此分层聚类更容易找到正确的聚类。

196. parallel ［ˈpærəlel］ adj. 平行的

例句 1　The three axes of the reference coordinate system are **parallel to** the three linear axes X, Y and Z of the machine tool coordinate system.

参考坐标系的三个轴平行于机床坐标系的三个直线轴 X、Y 和 Z。

例句 2　Two variants of the original pencil thrusters were developed **in parallel** to determine which would be the best solution for the team to utilize in executing a soft landing on the lunar surface.

平行开发了两个原始铅笔推进器，以确定哪个才是团队在月球表面软着陆时使用的最佳解决方案。

197. peak ［piːk］ n. 峰值

例句 1　The powder reflectivity has the greatest effect on the predicted **peak temperature** and melts pool geometry, followed by laser power and scanning speed.

粉末反射率对预测的峰值温度和熔池几何形状的影响最大，其次是激光功率和扫描速度。

例句 2　The **peak temperature** on machined surface in hard turning occurred along the intersection of cutting edge and the machined surface based on the prediction from the embedded thermocouple method.

根据嵌入式热电偶法的预测，硬车加工表面的峰值温度发生在切削刃与加工表面的交点处。

198. physical ['fɪzɪkl] adj. 物理的

例句 1　The advantage of such models is their relationship to the **physical process** of material removal and their ability to predict the jet footprint whenever the initial conditions are known.

这种模型的优点是它们与材料去除这一物理过程间的关系以及在已知初始条件下预测射流印迹的能力。

例句 2　The **physical constraints** include maximum chip thickness, cutting speed range for the machinability of the material, torque and power limits of the machine's tool and workpiece spindles, and chatter stability.

物理约束包括最大切屑厚度、材料可切削性切削速度范围、机床刀具和工件主轴的扭矩及功率限制，以及颤振稳定性。

199. plastic ['plæstɪk] adj. 塑性的

例句 1　Sub-surface **plastic deformation** was also seen in drilling Inconel 718 with minimum quantity lubrication (MQL).

在最小量润滑（MQL）的 Inconel 718 钻孔中也观察到了亚表面塑性变形。

例句 2　They include steep temperature gradients/rapid heating and quenching, chemical reactions and **plastic deformation**.

它们包括变化较大的温度梯度/快速加热和淬火，化学反应和塑性变形。

200. plot [plɒt] n. （表现两个变量关系的）图表

例句 1　The **contour plot** of a typical selected region from the surface along with measured surface statistics are depicted in Fig. 4.

图 4 描述了从表面选取的典型区域的等值线图以及测量得出的表面统计数据。

例句 2　When there is a trade-off between adjacent pairs of objectives, as in this illustration, the Pareto solution lines cross in the **parallel coordinate plot**.

当相邻目标对之间存在权衡时（如此图所示），帕累托解线在平行坐标图中相交。

201. plotted [plɔtɪd] v. 标绘 (过去分词)

例句 1　This parameter allows geometric stability of a wide range of frequencies to be **plotted in** one figure (Fig. 3).

该参数允许在一个图中绘制宽频率范围的几何稳定性（见图 3）。

例句 2　The deformation zones were revealed by backscattered electron (BSE) microscopy (Fig. 3) and the depths **plotted in** Fig. 4.

通过反向散射电子(BSE)显微镜(见图3)和图4中绘制的深度揭示了变形区。

202. polymer　[ˈpɑːlɪmər]　n. [高分子]聚合物

例句 1　Fig. 5 shows the modular designs we investigated in the **polymer prototype**.

图5显示了我们在聚合物原型中研究的模块化设计。

例句 2　In this paper, a test part is designed to evaluate the accuracy and resolution of the **polymer powder** bed fusion (PBF) or selective laser sintering process for a wide variety of features.

在本文中，测试部件的设计用来评估聚合物粉末床熔融(PBF)或选择性激光烧结工艺的各种特征的精度和分辨率。

203. porosity　[pɔːˈrɑːsəti]　n. 有孔性

例句 1　In addition, some processing defects such **as micro-porosity** and rough surface finish affects the final integrity of the produced part, resulting in parts with inconsistent performances.

此外，一些加工缺陷，如微孔和较大的表面粗糙度，会影响所生产零件的最终完整性，导致零件性能不一致。

例句 2　From an application stand point, triply periodic structures are gaining considerable interest for tissue engineering applications, particularly given such structures mimic the **porosity** of trabecular bone.

从应用的角度来看，三重周期性结构在组织工程应用方面获得了人们相当大的观注，特别是考虑到这些结构模仿小梁骨的孔隙率。

204. positive　[ˈpɑːzətɪv]　adj. 正向的

例句 1　This is a consequence of a **positive** down boundary shown in Fig. 5 where there is an unstable region for forces lower than the threshold value.

这是图5所示的正向下边界的结果，其中存在低于阈值的力的不稳定区域。

例句 2　The synergy between fixturing and AM is not limited to be used with complex geometries or to ensure a proper fixture/workpiece interaction, but can also have a **positive effect on** the manufacturing process itself.

夹具和增材制造之间的协同作用不仅不受复杂的几何形状或确保适当的夹具/

工件相互作用等要求所限，还可以对制造过程本身产生积极影响。

205. potential ［pəˈtenʃl］ n. 潜力；adj. 潜在的

例句1 Nevertheless, both proposed global and local corner smoothing techniques deliver nearly identical contouring performance, validating the **potential for** practical application of the developed technique.

尽管如此，所提出的全局和局部角平滑技术都提供了几乎相同的轮廓性能，验证了所开发技术的实际应用潜力。

例句2 The developed modelling framework provides the process design domain for different grinding process parameters, while maintaining a targeted wheel life, and averting **potential damage** of the workpiece.

开发的建模框架为不同的磨削工艺参数提供工艺设计域的同时，还保持了目标砂轮寿命，避免了工件的潜在损坏。

206. potentially ［pəˈtenʃəli］ adv. 潜在地

例句1 In order to improve the long-term stability, different thermal compensation techniques can be **potentially** implemented.

为了提高长期稳定性，可以采用不同的热补偿技术。

例句2 In addition to academic interests, such paradoxical mechanisms also could be **potentially** application-oriented once the mystery of their existences is revealed.

除了具有学术研究的价值之外，一旦它们存在的奥秘被揭开，这种自相矛盾的机制也可能具有潜在的应用导向性。

207. precision ［prɪˈsɪʒn］ n. 精度

例句1 One is the **set-up precision** of the spindle tool cup and the other is the position of the pivot tool cup.

一个是主轴刀具杯的设置精度，另一个是枢轴刀具杯的位置。

例句2 In material removal processes at **precision** and ultra-precision scales, the undeformed chip thickness can be in the order of a few microns or less and can even approach the nano scale.

在精密和超精密尺度的材料去除过程中，未变形的切屑厚度可以是几微米或更小，甚至可以接近纳米级。

208. predict ［prɪˈdɪkt］ v. 预测

例句 1　They extended the model to **predict** cutting forces for turning and drilling operations with indexable tools.

他们扩展了该模型，以预测使用可转位刀具进行车削和钻孔操作的切削力。

例句 2　The dynamics of metal cutting operations have also been extensively studied to **predict** forced vibrations and chatter stability both in frequency and discrete time domains.

金属切削操作的动力学也得到了广泛的研究，以预测频域和离散时域中的强迫振动和颤振稳定性。

209. predicted ［prɪˈdɪktɪd］ v. 预测（过去分词）

例句 1　The results indicated that roughness **predicted by** mathematical model has good agreement with the experimentally observed roughness.

结果表明，数学模型预测的粗糙度与实验观察到的粗糙度具有良好的一致性。

例句 2　Yet the higher stability at lower speeds is not **predicted by** the presented model due to neglecting the effect of high frequency torsional-axial vibrations on process damping.

然而，由于忽略了高频扭转轴向振动对过程阻尼的影响，因而所提出的模型无法预测较低速度下的较高稳定性。

210. prediction ［prɪˈdɪkʃn］ n. 预测

例句 1　Graphical techniques can also be employed for **prediction of** dynamic stability charts.

图形技术也可用于动态稳定性图表的预测。

例句 2　Although any tool geometry can be handled by the generalized model, a ball end mill is used to illustrate the **prediction of** the process mechanics.

虽然广义模型可以处理任何刀具的几何形状，但还是可以用球头立铣刀来阐释工艺力学的预测。

211. predictions ［prɪˈdɪkʃənz］ n. 预测（复数）

例句 1　The measured output of the actuator is in reasonable agreement with the **predictions of** the model.

执行器的输出测量值与模型预测值较好地吻合。

例句2　The speed-adapted **prediction** is generated by assembly of the predictions at the different spindle speeds.

速度自适应预测是由不同主轴转速下的预测组合生成的。

212. previously　[ˈpriːviəsli]　adv. 先前地

例句1　As the scan continues, the **previously** melted region begins to solidify due to heat loss from conduction and radiation.

随着扫描的继续，由于导热和辐射的热量损失，先前熔化的区域开始凝固。

例句2　Measurement results for relative permittivity, coercive field, and remnant polarization were found to be consistent with **previously** reported values.

相对介电常数、矫顽磁场和残余极化的测量结果与先前报道的结果一致。

213. primary　[ˈpraɪməri]　adj. 主要的

例句1　Here the field strength is shown to be larger in the **primary machining zone** extending into the central area.

这里显示，在延伸到中心区域的主加工区时，场强度更大。

例句2　Nevertheless, the **primary function** of the coolant is to reduce heat by decreasing the friction at the cutting zone.

然而，冷却剂的主要功能是通过减少切削区摩擦来降低热量。

214. prior　[ˈpraɪə(r)]　adj. 先前的，在前面的

例句1　For force prediction test a 17 mm hole is drilled **prior to** the operation.

为了进行力预测测试，需要在操作前钻一个17 mm的孔。

例句2　This section is used to summarize this **prior work** and provide support for the current investigation.

本节将对之前的工作进行总结，并为当下的研究提供支持。

215. procedure　[prəˈsiːdʒə(r)]　n. 程序

例句1　Next, the **procedure** is applied for cost modelling of spur gears.

接下来，将该程序应用于直齿圆柱齿轮的成本建模。

例句2　Furthermore, the simplicity of the **procedure** makes it an ideal method for real-time application.

此外，程序的简单性使其成为实时应用的理想方法。

216. processes [ˈprɒsesɪz] n. 过程，工艺(复数)

例句1 For example, the variability in surface orientation may prove to be a significant factor in deciding between **manufacturing processes**.

例如，表面取向的差异可能被证明是在两个制造过程中决定取舍的重要因素。

例句2 In contrast, additive manufacturing (AM) techniques do not have the same extent of design constraints that limit **conventional processes**.

相比之下，增材制造技术的设计约束程度并不像传统工艺那样大。

217. processing [ˈprəʊsesɪŋ] n. 加工，处理

例句1 The **processing** occurs out of the view of the camera for 99 frames.

该处理发生在99帧的镜头之外。

例句2 The effect of **processing parameters** on melt pool shape can be represented via main effect plots, as seen in Figs. 14-16.

加工参数对熔池形状的影响可以用主效应图表示，如图14至图16所示。

218. profile [ˈprəʊfaɪl] n. 轮廓，剖面

例句1 However, if the overlapping is large, the linear model is not accurate enough to predict the **profile**.

但是，如果重叠较大，线性模型就不足以准确地预测剖面。

例句2 **Surface profile** images of the device when the inner axis and outer axis are actuated as in Fig. 11.

驱动内、外轴时装置的表面轮廓图像如图11所示。

219. proportional [prəˈpɔːʃənl] adj. 成比例的

例句1 Work speed is **inversely proportional to** the number of lobes for a particular chatter frequency.

在某一颤振频率下，工作速度与波瓣数成反比。

例句2 Dynamic component of the ploughing force is **proportional to** the extruded workpiece volume under the tool flank.

犁削力的动态分量与刀具后刀面挤压工件的体积成正比。

220. radial [ˈreɪdiəl] adj. 径向的

例句1 The **radial** run-out of inserts are measured when the cutter was mounted on the spindle.

当刀具安装在主轴上时，可以测量刀片的径向跳动。

例句2　This can be calculated from the **radial** and axial immersion, tool diameter, feed rate and spindle speed for each tool move.

这可以从径向和轴向切入、刀具直径、每一个刀具移动的进给速度和主轴速度计算出来。

221. real-time　［ˌriːəl ˈtaɪm］　adj. 实时的

例句1　The main goal is to investigate whether these coefficients, estimated and tracked **in real-time**, can be used as an indirect method of estimating the tool's condition.

主要目标是研究这些实时估计和跟踪的系数是否可以作为判断刀具状态的间接方法。

例句2　However, utilizing iterative schemes for optimizing such a short motion is not feasible in **real-time implementation**, and should not improve the overall cycle time significantly.

但是，利用迭代方案对这类短动作进行优化在实时操作中是不可行的，并且不能显著优化整个周期时长。

222. region　［ˈriːdʒən］　n. 区域

例句1　In order to remedy this, a lens glare filter is implemented **in the region** close to the melt pool.

为了解决这个问题，在靠近熔池的区域安装了一个镜头眩光过滤器。

例句2　This methodology was employed to illustrate that our technique may be applied to any **region** of interest using any of the unit cells examined in this study for scaffold creation.

这种方法被用来说明我们的技术可以应用到任何感兴趣的区域，使用本研究中检查的任何单元格来创建支架。

223. regions　［ˈriːdʒənz］　n. 区域（复数）

例句1　This method treats all dimensions equally and is able to track the nonlinear **regions** in the input space.

该方法对所有维度一视同仁，能够跟踪输入空间的非线性区域。

例句 2 Over the course of numerous back and forth scanning, the solidified elevated and shallow **regions** will form wave-like structure.

在多次来回扫描的过程中，凝固的隆起和浅滩区域会形成波状结构。

224. removal [rɪˈmuːvl] n. 去除

例句 1 An abrasive total sliding length per unit **material removal rate** was established to assess the wheel radial wear.

建立了单位材料去除率的磨料总滑动长度来评估车轮径向磨损。

例句 2 In order to improve the surface quality of small core-drilled holes in composite materials, tools offering a more efficient **material removal mechanism** are required.

为了提高复合材料中小空心钻孔的表面质量，需要能够提供配备了更有效材料去除机制的刀具。

225. remove [rɪˈmuːv] v. 去除

例句 1 In manufacturing, machining can result in the need to **remove** unwanted material from the system.

在制造业中，机械加工就是从系统中去除不想要的材料。

例句 2 The chip thickness would be accumulated with subsequent rotation of the workpiece until the actual chip thickness approaches to the minimum chip thickness value to **remove** the material.

随着工件的转动，切屑厚度不断累积，直到实际切屑厚度接近最小切屑厚度值，才能去除材料。

226. removed [rɪˈmuːvd] v. 去除（过去分词）

例句 1 First, the quadratic model construction and local optimization steps are **removed**.

首先，去除二次模型构造和局部优化步骤。

例句 2 The shear plane is moving along the cutting direction during material removal and always connecting the grooved machined surface and the surface to be **removed**.

剪切面在材料去除过程中沿切削方向运动，始终连接开槽加工表面与待去除表面。

227. require ［rɪˈkwaɪə(r)］ v. 需要

例句1　The prediction of vibrations, hole surface quality and chatter-free spindle speed and pitch length **require** the dynamic model of orbital drilling.

振动、孔表面质量、无颤振主轴转速和螺距长度的预测需要建立轨道钻孔的动力学模型。

例句2　Today's highly complex designs **require** modern software tools and the realities of a global economy often constrain engineers to remote collaboration.

当今高度复杂的设计需要现代软件工具，而全球化经济的现实往往限制了工程师们的远程协作。

228. requirements ［rɪˈkwaɪəmənts］ n. 需求（复数）

例句1　The kinematic demands resulting from tool path shape also correspond to specific **energy requirements**.

由刀具路径形状产生的运动要求也对应于特定的能量需求。

例句2　When converting the topological graphs into functional linkages, additional efforts are needed to tackle the **design requirements**.

当将拓扑图转换为功能连接时，需要额外的工作来应对设计需求。

229. requires ［rɪˈkwaɪəz］ v. 需要（第三人称单数一般现在时）

例句1　The process of attaining the vacuum environment in the furnace **requires** multiple steps.

炉内达到真空环境的过程需要多个步骤。

例句2　This technique **requires** large computational effort, since it effectively moves feed rate selection from a preprocessing stage to the machine tool controller.

这种技术需要大量的计算工作，因为它有效地将进给速率的选择从预处理阶段转移到机床刀具控制器。

230. researchers ［ˈrɪsɜːtʃəz］ n. 研究人员（复数）

例句1　The paper presents a method for selecting grinding conditions and assists **researchers** to understand the complex dynamics of centreless grinding.

本文提出了一种选择磨削条件的方法，有助于研究人员理解无心磨削的复杂动力学过程。

例句2　Real signal monitoring using established thresholds is another method that has been studied by many **researchers** to detect tooth breakage.

使用已建立的阈值进行真实信号监测是许多研究人员研究的另一种检测轮齿破损的方法。

231. resolution ［ˌrezəˈluːʃn］ n. 清晰度，分辨率

例句 1　The high **resolution of** the profilometer revealed these small scale, high frequency features of the machined kerf.

轮廓仪的高分辨率揭示了加工切口的这些小尺度、高频特征。

例句 2　The second limitation greatly reduced the applicability of Box-Wilson central composite design types, which require high **resolution in** between levels.

第二个限制极大地降低了 Box-Wilson 中心复合设计类型的适用性，这需要在层次之间具有高分辨率。

232. response ［rɪˈspɒns］ n. 响应

例句 1　These can be measured from the **response** of any one resonator due to collective behaviour of the coupled system.

由于耦合系统的集体行为，这些可以从任何一个谐振器的响应中测量出来。

例句 2　The devices have also been tested with a laser Doppler vibrometer (LDV) to characterize **frequency response** and bandwidth.

该装置还经过了激光多普勒振动计的测试，以表征频率响应和带宽。

233. responses ［rɪˈspɒnsɪz］ n. 响应（复数）

例句 1　There is a need to select work speed in relation to geometry and machine **frequency responses**.

有必要根据几何形状和机器频率响应来选择工作速度。

例句 2　Neither the control group nor the group using the study showed a statistically or practically significant change in their **responses** in the pre/post-assessment.

对照组和使用该研究的组在测试前后评估的反应中都没有显示出统计学上或实际上的显著变化。

234. rigid ［ˈrɪdʒɪd］ adj. 刚性的

例句 1　Since the part is **rigid**, all the modal parameters of the tool represent the overall tool-workpiece system.

由于该零件是刚性的，刀具的所有模态参数都代表了整个刀具-工件系统。

例句 2　This location scheme can uniquely locate a **rigid** body without creating locator interference by using a minimum number of points in the following way.

该定位方案可以通过以下方式使用最小数量的点，在不产生定位器干扰的情况下独特地定位刚体。

235. robust ［rəʊˈbʌst］ adj. 稳健的，坚固的

例句 1　This result is comparable to the combined failure rate and shows that the effect size is **robust** to additional data.

这一结果与综合故障率相当，表明效应大小对附加数据是稳健的。

例句 2　Current methodologies cannot create complex surfaces in an efficient and scalable manner in **robust** engineering materials.

当前的方法无法在坚固的工程材料中以有效和可扩展的方式创建复杂的表面。

236. role ［rəʊl］ n. 作用

例句 1　This refutes prior work that found complexity to **play a role in** cognitive load.

这反驳了之前发现复杂性在认知负荷中起作用的研究。

例句 2　Finishing of the machined surface deteriorates with ploughing which also **plays a crucial role in** energy consumption.

机加工表面粗糙度会随着犁耕而下降，这在能源消耗中也起着至关重要的作用。

237. rotation ［rəʊˈteɪʃn］ n. 旋转

例句 1　For 5-axis milling, there would be additional lead and tilt angle **rotation transformations**.

五轴铣削时，会有额外的导程和倾角旋转变换。

例句 2　Ultimate fibre separation could have been as a result of tensile fracture caused by the **drill rotation**.

最终的纤维分离可能是由钻头旋转引起的拉伸断裂造成的。

238. rotational ［rəʊˈteɪʃnl］ adj. 旋转的

例句 1　A controller attempts to interpolate discretised tool paths by coordinating motion of independent translational and/or **rotational axes**.

控制器试图通过协调独立的直线轴和/或旋转轴的运动来对离散刀具路径进行插补。

例句2　This might be a possible reason for higher axis tracking error shown by X-axis and C-axis as compared to other translational and **rotational axes** respectively.

这可能是 X 轴和 C 轴相对于其他直线轴和旋转轴显示较大的轴跟踪误差的原因。

239. scan　[skæn]　n. 扫描

例句1　Fig. 3 illustrates this concept for both of these laser **scan strategies**.

图3解释了在这两种激光扫描策略中的这一概念。

例句2　Additionally **scan strategies** have been comprehensively studied and provide a myriad of combinations in order to alter the residual stresses in the as-built state.

另外，经过全面研究，扫描策略可以提供各种各样的组合，以改变竣工状态下的残余应力。

240. scanning　[ˈskænɪŋ]　adj. 扫描的

例句1　The powder reflectivity has the greatest effect on the predicted peak temperature and melts pool geometry, followed by laser power and **scanning speed**.

粉末反射率对预测峰值温度和熔体熔池几何形状的影响最大，其次是激光功率和扫描速度。

例句2　Depth profiling, digital microscopy and **scanning electron microscopy** are utilized to investigate topological evolution and mechanisms of grit failure.

深度剖析、数字显微镜和扫描电子显微镜用于研究磨粒破坏的拓扑演化及机制。

241. sections　[ˈsekʃnz]　n. 部分（复数）

例句1　The **sections of** the thruster would then need to be mated together with a strong sealing interface during assembly.

在装配过程中，推进器的各个部分需要与一个坚固的密封接口连接。

例句2　The **following sections** outline the approaches to minimise the set-up errors in the spindle tool cup and the centre pivot tool cup.

以下部分概述了将主轴刀杯和中心枢轴刀杯的安装误差降到最低的方法。

242. segments ['seɡmənts] n. 片段；段数(复数)

例句1 As shown, when corners are locally smoothened by proposed technique from Section 2, the path consists of **linear segments** and corner transitions.

如图所示，当使用第2节中提出的技术对角点进行局部平滑时，路径由线段和角点过渡组成。

例句2 Turning and boring tools are radially immersed into the workpiece, thus the **cutting edge segments** are aligned with the radial direction of the workpiece.

将车刀和镗刀径向切入工件中，使切削刃段与工件径向方向对齐。

243. selected [sɪ'lektɪd] v. 选择(过去分词)

例句1 The **selected area** diffraction patterns in Fig. 7 are more continuous for the wet as compared to dry cutting suggesting less lattice rotation in dry cutting.

与干切削相比，图7中选定的区域衍射模式在湿切削时更连续，这表明在干切削时晶格旋转更少。

例句2 Six out of the 160 poses of the measurement cycle trajectory were **selected** for further analysis because of the large errors observed at those poses.

在测量周期轨迹的160个位姿中，选择了6个位姿进行进一步分析，因为在这些位姿下观测到了较大误差。

244. selection [sɪ'lekʃn] n. 选择

例句1 Practical implications are explained for **selection of** grinding conditions.

阐述了磨削条件选择的实际意义。

例句2 The **selection of** the process parameters is illustrated for a turn milling example where the tool is a two fluted end mill having 16 mm diameter.

工艺参数的选择以车铣为例进行说明，其中刀具是直径为16 mm的双槽立铣刀。

245. sensing ['sensɪŋ] n. 传感，感应

例句1 Changes in the measurand can be considered as external perturbations to the system in **sensing** applications.

在传感应用中，测量值的变化可以被认为是对系统的外部扰动。

例句2 This **sensing** will be a cornerstone of integrating "cloud" and "Internet-of-things" functionally into the next generation manufacturing environment.

这种感应将成为将"云"和"物联网"功能集成到下一代制造环境中的基石。

246. sensors ['sensəz] n. 传感器(复数)

例句 1 The distances to all **sensors** are used to determine the position of the spindle ball in the Capball frame via a least squares sphere algorithm.

通过最小二乘球面算法,所有传感器之间的距离被用来确定 Capball 框架中主轴球的位置。

例句 2 The sensitivity of pressure **sensors** decreased at high temperature due to the decrease of the gauge factor of SiC at elevated temperatures, as presented in the previous section.

如前一节所述,压力传感器的灵敏度在高温下降低,这是因为高温下碳化硅的灵敏系数降低了。

247. series ['sɪəriːz] n. 系列

例句 1 The system works through **a series of** half cylinders, with their flat surface acting as the clamping elements.

该系统通过一系列半圆柱体工作,半圆柱体的平面用作夹紧元件。

例句 2 Innovative location and/or holding technologies represents **a series of** inventive hardware solutions designed to address different aspects of fixturing such as flexibility and the holding of intricate geometries.

创新的定位和/或夹持技术代表了一系列创造性的硬件解决方案,旨在解决夹具的不同方面,例如灵活性和复杂几何形状的夹持。

248. setup ['setʌp] n. 装置,安装,建立

例句 1 However, these measurement systems are expensive and the **setup of** the instrument is time-consuming.

然而,这些测量系统昂贵,而且仪器安装费时。

例句 2 The hard turning **experimental setup** and process parameters used in this study are first shown in Section 2.

首先在第 2 节中展示本研究中使用的硬车削实验装置和工艺参数。

249. similarly ['sɪmələli] adv. 相似地,同样地,也

例句 1 The frequency drift of the high-temperature pressure sensor is determined similarly to that of the proof-of-concept pressure sensor.

高温压力传感器的频率漂移确定方法与概念验证式压力传感器的确定相似。

例句 2　It is noticed that the larger grain material produces smaller machining forces than that of smaller grain materials **similarly** observed in micromachining.

值得注意的是，在微加工中类似地观察到，较大晶粒材料产生的加工力小于较小晶粒材料。

250. simulated　['sɪmjuleɪtɪd]　v. 模拟，模仿(过去分词)

例句 1　Temperature rise due to two heat sources is **simulated** independently based on such model setup.

在此模型建立的基础上，对两个热源引起的温度升高进行了独立的模拟。

例句 2　The model **simulated** brittle erosion using two damage mechanisms: crater removal due to lateral crack formation and edge chipping.

该模型采用两种损伤机制模拟脆性侵蚀：侧向裂纹形成导致的陨坑清除和尖锐切屑。

251. simulations　[ˌsɪmju'leɪʃnz]　n. 模拟，仿真(复数)

例句 1　Significant prior work has been carried out in the development and deployment of **simulations of** energy beam processes.

在能量束过程模拟的开发和部署方面已经开展了重要的前期工作。

例句 2　The **simulations** are validated by measured cutting forces of indexable drilling and milling cutters, multifunctional drilling tool, serrated end mill and cylindrical end mill.

仿真结果通过可转位钻铣刀、多功能钻铣刀、锯齿立铣刀和外圆立铣刀的切削力测试进行验证。

252. simultaneously　[ˌsɪm(ə)l'teɪnɪəsli]　adv. 同时地

例句 1　Furthermore, their study is unique in overcoming these three fundamental problems **simultaneously**.

此外，他们研究的独特之处在于能同时克服这三个基本问题。

例句 2　It is common to come across situations in engineering design where multiple conflicting performance criteria need to be **simultaneously** optimized.

在工程设计中经常会遇到需要同时优化多个相互冲突的性能标准的情况。

253. software ['sɒftweə(r)] n. 软件

例句1 To manufacture computer-aided design (CAD) models, computer-aided manufacturing (CAM) **software** can produce commands for computer numerically controlled (CNC) machines.

为了加工计算机辅助设计模型，计算机辅助制造软件可以为计算机数控机床生成指令。

例句2 However, several **software-assisted methods** have been developed for various aspects of design for manufacturing and process selection.

然而，已经针对制造设计和工艺选择的各个方面开发了一些软件辅助方法。

254. specific [spəˈsɪfɪk] adj. 具体的

例句1 The **specific values** of feed rate and line length chosen are immaterial.

所选的进料速率和线长度的具体值并不重要。

例句2 The latter has included **specific effort** towards the smoothing of tool axis orientation.

后者包括对平滑刀轴方向的具体尝试。

255. specifically [spəˈsɪfɪkli] adv. 具体地，特别地

例句1 **Specifically**, a tetrahedron is used in this paper.

具体来说，本文中使用了一个四面体。

例句2 An inertial sensor, **specifically** a tri-axial accelerometer is used as an independent source of measurement.

惯性传感器，特别是三轴加速度计被用作一个独立的测量源。

256. specified [ˈspesɪfaɪd] v. 特定，指定（过去分词）

例句1 If these requirements exceed physical capabilities of motors, **specified** motions may be compromised.

如果这些要求超出了电动机的物理性能，特定的运动可能会受到影响。

例句2 In the type synthesis process, each exoskeleton is derived from a **specified** driving four bar linkage.

在类型合成过程中，每个外骨架都来自指定的驱动四杆机构。

257. stability ［stə'bɪləti］ n. 稳定性

例句 1 A dyn is termed the dynamic **stability** parameter.

动态稳定性参数被称为 dyn。

例句 2 This is a problem for centreless grinding where there is particular interest in **stability** of integer lobes.

这是无心磨削中的一个问题，其中特别关注的是整数叶的稳定性。

258. stable ［'steɪbl］ adj. 稳定的

例句 1 For square shape once the grit is in full contact with the workpiece the ploughing action is **stable**.

一旦方形磨粒与工件充分接触，犁削作用是稳定的。

例句 2 In the framework of statistical quality control, this natural variability is often called a **stable system** of chance causes.

在统计质量控制的框架中，这种自然的可变性通常被称为偶然原因的稳定系统。

259. static ［'stætɪk］ adj. 静态的

例句 1 These models are all based on the assumption of a **static**, cylindrical Gaussian energy distribution and have produced reliable results.

这些模型都是建立在静态的圆柱形高斯能量分布的假设基础上，并得出了可靠的结果。

例句 2 These induced lateral deformations could significantly influence **static** load distribution, leading to inaccurate predictions by such models.

这些诱发的横向变形可显著影响静态载荷分布，并导致通过这些模型预测不准确。

260. strain ［streɪn］ n. 应变，形变

例句 1 The level of **strain hardening** and the extent of zone B increase significantly with increasing feed rate.

应变硬化程度和 B 区范围会随着进给量的增加而明显增大。

例句 2 If the stress drop is larger than the effective stress, then the initial **plastic strain rate** should be negative (see Fig. 5).

如果应力降大于有效应力，则初始塑性应变率为负值(见图 5)。

261. strategy ['strætədʒi] n. 策略

例句 1　The traditional **strategy for** reducing error is to concentrate upon the minimization of position error for each axis.

减少误差的传统策略专注于每个轴的位置误差最小化。

例句 2　The **strategy mapping tool** presented in this work is a binary decision tree with different paths or set of nodes that lead to a solution.

本研究中提出的策略映射工具是一种具有不同路径或节点集的二叉决策树,可用于求解。

262. stress [stres] n. 应力

例句 1　In this respect, the **material flow stress** can be used to analyse the performance of the cutting process.

在这方面,材料流动应力可以用来分析切削过程的性能。

例句 2　The failure normal and **shear stress** are calculated from the maximum measured forces divided by the grit impression area.

故障正应力和剪应力是根据测得的最大力除以磨粒压痕面积计算出来的。

263. structural ['strʌktʃərəl] adj. 结构的

例句 1　Cutting forces excite the **structural dynamics** of the tool and part.

切削力会激发刀具和零件的结构动力。

例句 2　As a result, the **structural complexity** has strong impact on the effort (and cost) of system design, development, and operation.

因此,结构复杂性对系统设计、开发和运行的工作量(以及成本)有很大的影响。

264. structures [ˈstrʌktʃəz] n. 结构(复数)

例句 1　Rounding problems are minimised by **stiff structures**, low work speeds and high damping.

刚性结构、低工作速度和高阻尼可以最大限度地减少倒圆问题。

例句 2　Future studies could add functional features such as **compliant structures**, gears, and springs.

未来的研究可以增加一些功能特征,如顺应结构、齿轮和弹簧。

265. substrate ['sʌbstreɪt] n. 基板，基底

例句1　One of the arms was connected to the paddle and the end of the other arm was connected to the **substrate** via the anchor.

其中一臂连接到桨叶，另一臂的一端通过锚点连接到基板。

例句2　On the other hand, large melt pools may yield vaporization of the **substrate** and causes pores in the structure that increase porosity.

另一方面，大的熔池可能会产生基底汽化，并导致在孔隙率增加的结构中产生孔隙。

266. sufficient [sə'fɪʃnt] adj. 充分的，足够的

例句1　For example, motors must provide **sufficient power** to overcome machine inertia, cutting forces and friction.

例如，电动机必须提供足够的动力来克服机器惯性、切削力和摩擦。

例句2　As a result, the affected drill degrades rapidly without **sufficient cooling** and lubrication and its ultimate tool life will be deeply impacted.

因此，受影响的钻头在没有充分冷却和润滑的情况下迅速退化，并将严重影响刀具最终寿命。

267. summarized ['sʌməraɪzd] v. 总结，概括（过去分词）

例句1　The test procedure is **summarized on** a flow chart in Fig. 2.

试验程序被总结在图2的流程图里。

例句2　Temperature dependent material properties are **summarized in** Fig. 3.

图3中对材料随温度变化的特性进行了总结。

268. tangential [tæn'dʒenʃl] adj. 切线的，切向的

例句1　The **tangential and radial force**（Fig. 1）both consist of a shearing component and a rubbing component.

切向力和径向力（图1）均由一个剪切分量和一个摩擦分量组成。

例句2　Or alternatively, if the specific energy for chip formation is known, the **tangential forces** could be predicted.

另外，如果已知切屑形成的比能，则可以预测切向力。

269. target ['tɑːgɪt] n. 目标

例句1　In other words, the **target surface** is usually defined without checking whether the process can etch that particular surface.

换句话说，在定义目标表面时，通常不对此工艺是否能蚀刻该特定表面进行检查。

例句2　The tracking error between the desired trajectory and the **target** one is reduced and the quality of the etched surface is improved.

期望轨迹与目标轨迹之间的跟踪误差减小了，同时蚀刻表面的质量提高了。

270. task [tɑːsk] n. 任务

例句1　It depends on the nature of the **task** and assembly product design.

它取决于任务的性质和装配产品设计。

例句2　In contrast, the **task** in the dirt-bike tire changing tool challenge was to design a tire changing tool.

相比之下，越野摩托车换胎工具挑战的任务就是设计一个换胎工具。

271. tasks [tɑːsks] n. 任务（复数）

例句1　The **tasks** in both studies can be classified as cases of full configuration design.

这两项研究的任务可以归类为全配置设计的案例。

例句2　Future research may also focus on replicating the study with professional designers using more complex **design tasks**.

未来的研究可能还会集中于用专业设计师和更复杂的设计任务来复制此研究。

272. technologies [tekˈnɒlədʒiz] n. 技术（复数）

例句1　Hence, there is a need for advanced **manufacturing technologies** which overcome this.

因此，需要采用先进的制造技术克服这一问题。

例句2　These **technologies** are mostly dedicated to produce a particular material classification, polymer, metal, ceramics, or composites.

这些技术大多致力于生产特定的材料类别，如聚合物、金属、陶瓷或复合材料。

273. technology [tekˈnɒlədʒi] n. 技术

例句1　Such a study could be done on a new product or **existing technology**.

这种研究可以在新产品或现有技术上进行。

例句2　The parameters in the models are estimated based on **past technology** performance data.

模型中的参数是根据以往技术性能数据来估计的。

274. theoretical [ˌθɪəˈretɪkl] adj. 理论的

例句1　The measured frequencies matched well with the **theoretical calculations**.

实测频率与理论计算结果吻合较好。

例句2　The actual curvature at which this transition begins must therefore be less than the **theoretical** transition curvature.

因此，过渡开始时的实际曲率必须小于理论上的过渡曲率。

275. theory [ˈθɪəri] n. 理论

例句1　**In the theory**, only one trench is necessary for calibration purposes.

理论上，校准只需一个沟槽。

例句2　**In the theory of** highly optimized tolerance, an increase of part count especially the number of unique parts escalates system robustness.

在高度优化公差理论中，零件数量的增加，尤其是独特零件数量的增加，可以提高系统的鲁棒性。

276. thermal [ˈθɜːm(ə)l] adj. 热的，热量的

例句1　This work has been augmented by studies using high-speed **thermal imaging** of the two tool types in operation.

通过对两种工具的高速热成像研究，这项研究工作得以加强。

例句2　Machining of nickel alloys is difficult due to a combination of high temperature strength and low **thermal conductivity**.

由于高温强度和低导热系数的综合作用，镍合金的加工很难进行。

277. threshold [ˈθreʃhəʊld] n. 阈值

例句1　Positive up boundaries indicate instability for forces larger than the **threshold value**.

正向上边界显示大于阈值的力的不稳定性。

例句 2　As in the hierarchical clustering, a **threshold** is needed to generate clusters from the dendrogram.

与分层聚类一样，从树状图生成聚类需要一个阈值。

278. tissue　[ˈtɪʃuː]　n. 组织

例句 1　Due to the asymmetry, the **tissue** is displaced by the needle tip as the needle cuts through tissue.

由于不对称，当针穿过组织时，组织会被针尖移位。

例句 2　This procedure was chosen due to its demanding requirements for stability, dexterity, and force application to **tissue**.

选择这一流程是由于其对稳定性、灵活性和组织应用力的要求。

279. traditional　[trəˈdɪʃənl]　adj. 传统的

例句 1　The **traditional** strategy for reducing error is to concentrate upon the minimization of position error for each axis.

减少误差的传统策略专注于每个轴的位置误差的最小化。

例句 2　Both the watts and the **traditional** numerical scale indicate the degree to which the toast will be cooked, the "doneness".

瓦数和传统的数字刻度都表示出面包的烘烤程度，即"熟度"。

280. transfer　[trænsˈfɜː(r)]　n. 传递，传导

例句 1　The oscillations in the measured forces are due to the **transfer function** of the rotary force dynamometer.

所测力的振荡是由回转式测力仪的传递函数引起的。

例句 2　This method requires the development of inverse heat **transfer method** and numerical thermal model for hard turning to predict the peak machined surface temperature.

本方法需要开发硬车削的逆传热方法和数值热模型来预测加工表面的峰值温度。

281. transformation　[ˌtrænsfəˈmeɪʃn]　n. 变化，变换

例句 1　Therefore, in this paper the **coordinate transformation method** is employed for calculating the misalignment.

因此，本文采用坐标变换的方法计算误差。

例句2　Once the workpiece temperature exceeds the austenitization temperature, martensitic **phase transformation** takes place and leads to thermally induced white layers.

一旦工件温度超过奥氏体化温度，就会发生马氏体相变，并产生热致白层。

282. transient　['trænziənt]　n. 瞬态

例句1　This reduces **transient overshoots** but also leads to longer path error settling time.

这减少了瞬态过冲，但也导致更长的路径误差校正时间。

例句2　For this purpose, models were necessary to calculate the **transient system responses** of both devices for different accelerations loads.

为此，需要建立模型来计算两种装置在不同加速度载荷下的瞬态系统响应。

283. transition　[træn'zɪʃn]　n. 转变，转换

例句1　Therefore, finding the eigenvalues for the constant **state transition matrix** is sufficient.

因此，求出恒定状态下转换矩阵的特征值就足够了。

例句2　During the **phase transition**, thermal expansion and phase transition mechanisms coexist and produce displacements in different directions.

在相变过程中，热膨胀和相变机制共存，并产生不同方向的位移。

284. trend　[trend]　n. 趋势

例句1　The **trend lines** show that melt pool size increases slightly as energy density increases.

趋势线显示，随着能量密度的增加，熔池尺寸略有增大。

例句2　It is observed that the profile along the transverse direction shows **similar trend** in all the cases.

可以看出，在所有情况下，沿横向的剖面都显示出类似的趋势。

285. uniform　['juːnɪfɔːm]　adj. 均匀的

例句1　Therefore, for very high beam speed, the melt flow may not be sufficient to form a **uniform surface**.

因此，对于非常高的波束速度，熔体流动可能不足以形成均匀表面。

例句2　For the **uniform wear theory**, the wear rate of the lining material is assumed to be uniform over the frictional surface.

均匀磨损理论假设衬里材料的磨损速率在摩擦表面上是均匀的。

286. utilized　[ˈjuːtəlaɪzd]　v. 采用，利用（过去分词）

例句1　Nonetheless, various methods are **utilized to** monitor milling tool condition.

尽管如此，各种方法仍被用于监测铣削刀具的状态。

例句2　A single grit pullout device is developed and **utilized for** this analysis.

研发了单磨粒脱离装置并用于此分析。

287. vacuum　[ˈvækjuːm]　adj. 真空的；n. 真空

例句1　The process of attaining the **vacuum environment** in the furnace requires multiple steps.

炉内达到真空环境的过程需要多个步骤。

例句2　In fact, the beam equations ignore all damping and are only appropriate for describing the beam **in a vacuum**.

事实上，射束方程忽略了所有阻尼，只适合于描述真空中的波束。

288. validate　[ˈvælɪdeɪt]　v. 验证，证实

例句1　We used two different approaches to **validate** the worksheet.

我们使用两种不同的方法来验证工作表。

例句2　These results indirectly **validate** the accuracy of the measurement results.

这些结果间接地验证了测量结果的准确性。

289. validated　[ˈvælɪˌdeɪtid]　v. 验证，证实（过去分词）

例句1　The proposed method is **validated in** milling **tests**.

该建议方法在铣削试验中得到了验证。

例句2　The mechanics, stability and surface location error models have been **experimentally validated by** conducting orbital drilling of holes.

这些力学、稳定性和表面定位误差模型已经通过轨道钻孔实验得到了验证。

290. validation　[ˌvælɪˈdeɪʃn]　n. 验证

例句1　The measured deflection is later used as ground truth for **experimental vali-**

dation of the proposed model.

测得的挠度稍后被用作实验,验证该建议模型的地面真实值。

例句2　Both theoretical proof and **experimental validation** of stability of the new torque controller have been carried out and reported in this paper.

本文对该新型转矩控制器的稳定性进行了理论验证和实验验证。

291. variable ['veəriəbl] adj. 不同的,可变的；n. 变量

例句1　The simulation demonstrated that the resultant profile was found to be **variable** from that of a standard nozzle.

仿真结果表明,所得的截面形状与标准喷管的截面形状不同。

例句2　Accordingly, the validity of using the energy density **variable** as a means of process characterisation has recently been questioned.

因而,将能量密度变量用作过程表征的有效性近来受到质疑。

292. variation [ˌveəriˈeɪʃn] n. 变化

例句1　The stochastic nature of the simulation can be seen by the **variation in** the repeated profiles.

通过重复剖面的变化可以看出模拟的随机性。

例句2　The **variation of** the machined surface integrity is influenced by the tool edge radius effect in ultra-precision machining.

在超精密加工中,刀刃半径效应影响加工表面完整性的变化。

293. variations [ˌveəriːˈeiʃnz] n. 变化(复数)

例句1　Random fabrication **variations in** dimensions led to an intrinsic imbalance in the system.

随机制造尺寸的变化导致了系统内在的不平衡。

例句2　Cutting tool edge radius effect could be one of the factors causing the **variations of** the machining results.

刀刃半径效应可能是造成加工结果变化的因素之一。

294. vary ['veəri] v. (使)不同,变化

例句1　It can be seen that the force values **vary in response to** the changes in cut geometry.

可以看出，力值随切削几何形状的变化而变化。

例句2　Note that the cutting force coefficients may **vary** at different cutting speeds and chip thicknesses.

需要注意的是，切削力系数可能会因不同的切削速度和切屑厚度而不同。

295. varying　［ˈveərɪŋ］　v. 变化（现在分词或动名词）

例句1　Our models were validated through insertion experiments carried out into phantom tissue samples under **varying** experimental conditions.

通过在不同实验条件下对模拟组织样本进行插入实验，可以验证我们的模型。

例句2　In other words, **varying** a single or combinations of these drill geometries will influence the directions of coolant flow.

换句话说，改变这些钻头中的一个几何形状或它们的组合将影响冷却剂流动的方向。

296. vectors　［ˈvɛktəz］　n. 矢量，向量（复数）

例句1　The model starts by defining the tangent and rake face **vectors** at discrete elements along the cutting edge.

本模型首先对沿切削刃的离散元素的切线和前刀面矢量进行定义。

例句2　The modelling of the cutting edge location and its rake face orientation **vectors** for solid and indexable tools with arbitrary geometry is presented in Section 2.

第2节介绍了任意几何形状的实体和可转位刀具的切削刃位置及其前刀面方向矢量的建模。

297. velocities　［vəˈlɑsətiz］　n. 速度（复数）

例句1　We here consider the absolute linear **velocities** of the robot and the human.

这里我们考虑的是机器人和人的绝对线速度。

例句2　Achieved axes **velocities** and displacements are derived from axis acceleration profiles.

得到的轴速度和位移是由轴加速度曲线推导出来的。

298. versus　［ˈvɜːsəs］　prep. 与……相比

例句1　The weighting of energy **versus** comfort in the reward function can make a significant difference to the performance of the resulting policy.

在奖励函数中，能量与舒适度的权重会对结果策略的绩效产生显著的影响。

例句2 Leak checks were conducted to make sure the pressure **versus** flow rate data reflected what was occurring inside the compression chamber.

进行泄漏检查的目的在于确保压力与流量数据相比反映了压缩室内部发生的情况。

299. vertical ［ˈvɜːtɪkl］ adj. 垂直的

例句1 The actuator is oriented such that the jet emanating from it is along the **vertical axis**.

执行器被定向，以使它发出的射流沿着垂直轴方向。

例句2 This guaranteed that the recorded **vertical displacement** of the mirror platform corresponded to the outer hysteresis loop of each leg.

这保证了镜台记录的垂直位移与每个立柱的外滞回线相对应。

300. via ［ˈvaɪə］ prep. 通过，经由

例句1 This concept was expanded upon **via** an intelligent fixture with sensory feedback.

这个概念是通过一个具有感官反馈的智能夹具扩展得到的。

例句2 Solving the optimization problem **via** evolutionary algorithms returns the optimal values for the design variables.

通过进化算法求解优化问题可返回设计变量的最优值。

301. vibration ［vaɪˈbreɪʃn］ n. 振动

例句1 Note that in most cases the **vibration sensitivity** gets larger when the control loop is enabled.

注意在大多数情况下，当控制回路被启用时振动灵敏度会变大。

例句2 **Vibration instability**, known as regenerative chatter in machining processes, is caused by the feedback between the vibration waves generated in subsequent cuts.

振动不稳定，亦被称为机械加工过程中的再生颤振，是由后续切削过程中产生的振动波之间的反馈引起的。

302. whereas ［ˌweərˈæz］ conj. 然而

例句1 Lateral forces are predicted within 20% error, **whereas** axial force is predicted within 10% error.

侧向力预测误差在20%以内,而轴向力预测误差则在10%以内。

例句2　In the argon mode, many enclosed pores exist within the grain, **whereas** the sample sintered in the vacuum has almost no pores.

氩气模式下,晶粒内部有许多封闭的气孔,而真空烧结的样品几乎没有气孔。

References（参考文献）

[1] ABDELALL E, FRANK M C, Stone R. A study of design fixation related to Additive Manufacturing[J]. Journal of Mechanical Design, 2018, 140(4).

[2] ABDELALL E S, FRANK M C, Stone R T. Design for manufacturability-based feedback to mitigate design fixation[J]. Journal of Mechanical Design, 2018, 140(9): 091701.

[3] AHMADI K, SAVILOV A. modelling the mechanics and dynamics of arbitrary edge drills[J]. International Journal of Machine Tools & Manufacture, 2015, 89: 208-220.

[4] MARVIN A, NICHOLAS H, JENSEN D C. Exploring natural strategies for bio-inspired fault adaptive systems design[J]. Journal of Mechanical Design, 2018, 140(9): 091101.

[5] AXINTE D, BUTLER-SMITH P, AKGUN C, et al. On the influence of single grit micro-geometry on grinding behavior of ductile and brittle materials[J]. International Journal of Machine Tools & Manufacture, 2013, 74: 12-18.

[6] BOROUMAND AZAD J, REZADAD I, PEALE R E, et al. Ultraviolet-assisted release of microelectromechanical systems from polyimide sacrificial layer[J]. Journal of Microelectromechanical Systems, 2015, 24(6): 2027-2032.

[7] CHAKLADER R, PARKINSON M B. Data-driven sizing specification utilizing consumer text reviews[J]. Journal of Mechanical Design, 2017, 139(11): 111406.

[8] CHEN L, TAI B L, CHAUDHARI R G, et al. Machined surface temperature in hard turning[J]. International Journal of Machine Tools and Manufacture, 2017: S0890695516304813.

[9] CHENG G H, TIMOTHY G, GARY W G. An adaptive aggregation-based approach for expensively constrained black-box optimization problems[J]. Journal of Mechanical Design, 2018, 140(9): 091402.

[10] COMAK A, ALTINTAS Y. Mechanics of turn-milling operations[J]. International Journal of Machine Tools and Manufacture, 2017: 2-9.

[11] CRIALES L E, ARISOY Y M, LANE B, et al. Laser powder bed fusion of nickel alloy 625: experimental investigations of effects of process parameters on melt pool size and shape with spatter analysis[J]. International Journal of Machine Tools and Manufacture, 2017: S0890695516303108.

[12] TRUONG, DO, PATRICK, et al. Process development toward full-density stainless steel parts with binder jetting printing[J]. International Journal of Machine Tools & Manufacture, 2017.

[13] ELGENEIDY K, LOHSE N, JACKSON M R. Bending angle prediction and control of soft pneumatic actuators with embedded flex sensors - A data-driven approach[J]. Mechatronics, 2017.

[14] GUILLERNA, BILBAO A, AXINTE D, et al. The linear inverse problem in energy beam processing with an application to abrasive waterjet machining[J]. International Journal of Machine Tools & Manufacture: Design, research and application, 2015, 99: 34-42.

[15] HOEFER, MICHAEL J, MATTHEW C F. et al. Automated manufacturing process selection during conceptual design[J]. Journal of Mechanical Design, 2018, 140(3).

[16] JIANG X, CRIPPS R J. A method of testing position independent geometric errors in rotary axes of a five-axis machine tool using a double ball bar[J]. International Journal of Machine Tools & Manufacture, 2015, 89: 151-158.

[17] JIANG B, LAN S, NI J. modelling and experimental investigation of gas film in micro-electrochemical discharge machining process[J]. International Journal of Machine Tools and Manufacture, 2015, 90: 8-15.

[18] LAUFF C, KOTYS-SCHWARTZ D, RENTSCHLER M E. What is a prototype? What are the roles of prototypes in companies? [J]. Journal of Mechanical Design, 2018.

[19] NOURI M, FUSSELL B K, ZINITI B L, et al. Real-time tool wear monitoring in milling using a cutting condition independent method[J]. International Journal of Machine Tools and Manufacture, 2015, 89: 1-13.

[20] RAHAMAN M, SEETHALER R, YELLOWLEY I. A new approach to contour er-

ror control in high speed machining[J]. International Journal of Machine Tools & Manufacture, 2015, 88: 42-50.

[21] ROGERS, JOHN E, YOON Y K, et al. A passive wireless microelectromechanical pressure sensor for harsh environments[J]. Journal of Microelectromechanical Systems, 2017, 27(1): 73-85.

[22] ROWE B W. Rounding and stability in centreless grinding[J]. International Journal of Machine Tools & Manufacture, 2014, 82-83: 1-10.

[23] SCHNELLE S, WANG J, JAGACINSKI R, et al. A feedforward and feedback integrated lateral and longitudinal driver model for personalized advanced driver assistance systems[J]. Mechatronics, 2018, 50: 177-188.

[24] SHE J, MACDONALD E F. Exploring the effects of a product's sustainability triggers on pro-environmental decision-making[J]. Journal of Mechnaical Design, 2018, 140: 177-188.

[25] SHIPLEY H, MCDONNELL D, CULLETON M. et al. Optimisation of process parameters to address fundamental challenges during selective laser melting of Ti-6Al-4V: A review[J]. International Journal of Machine Tools & Manufacture: Design, research and application, 2018, 128: 1-20.

[26] SHRESTHA S, CHOU K. A build surface study of powder-bed electron beam additive manufacturing by 3D thermo-fluid simulation and white-light interferometry [J]. International Journal of Machine Tools and Manufacture, 2017, 121: 37-49.

[27] TAJIMA, SHINGO, SENCER, et al. Global tool-path smoothing for CNC machine tools with uninterrupted acceleration[J]. International Journal of Machine Tools & Manufacture: Design, research and application, 2017, 121: 81-95.

[28] UNAL M, WARN G P, SIMPSON T W. Quantifying tradeoffs to reduce the dimensionality of complex design optimization problems and expedite trade space exploration[J]. Structural and Multidisciplinary Optimization, 2016, 54(2): 1-16.

[29] 张展. 英汉机械工程常用词汇[M]. 北京: 机械工业出版社, 2017.